Fish Biology

Fish Biology

Connor Harper

R CALLISTO REFERENCE

www.callistoreference.com

Callisto Reference,
118-35 Queens Blvd., Suite 400,
Forest Hills, NY 11375, USA

Visit us on the World Wide Web at:
www.callistoreference.com

ISBN: 978-1-64116-466-5 (Hardback)

Cataloging-in-Publication Data

Fish biology / Connor Harper.
    p. cm.
Includes bibliographical references and index.
ISBN 978-1-64116-466-5
1. Fishes. 2. Fishes--Life cycles. 3. Fishes--Habitat. 4. Fishes--Ecology.
5. Fisheries. I. Harper, Connor.
QL615 .F57 2020
597--dc23

Table of Contents

Chapter 1

Ichthyology: An Introduction

The aquatic craniate animals which bear gills and do not have limbs with digits are called fish. The branch of zoology which focuses on the study of fish is known as ichthyology. This is an introductory chapter which will introduce briefly all the significant aspects related to fish and ichthyology.

A fish is a poikilothermic (cold-blooded), water dwelling vertebrate with gills throughout life, and limbs—if present—in the form of fins. Poikilothermic refers to the fact that the internal temperatures of fish vary, often matching the ambient temperature of the environment.

There are over 27,000 known extant (living) fish species, making them the most diverse group of vertebrates, with more than one-half of the total vertebrate species. A heterogeneous assemblage, modern fish are conventionally divided into the jawless fish (class or superclass Agnatha, about 75 species including lampreys and hagfish), the cartilaginous fish (class Chondrichthyes, about 800 species including sharks and rays), and the bony fish (class Osteichthyes, with over 26,000 species). Some individuals only use the term fish when referring to the jawed bony fish, and do not include Agnatha or Chondrichthyes.

Since the dawn of humanity, people and fish have been linked. Fish provide humans much of their protein, offer recreational use via fishing, provide a sense of beauty as ornamental fish, and even serve in religious symbolism. The relationship has not been as favorable for fish: in 1999, seventy percent of the world's major fish species were determined to be fully- or over-exploited.

The study of fish is called ichthyology.

Characteristics of Fish

Fish range in size from the 14m (45ft) whale shark to the 7mm (just over 1/4 of an inch) long stout infantfish and the 13mm Philippine goby. Fish can be found in almost all large bodies of water in salt, or brackish, or fresh water, at depths from just below the surface to several thousand meters. However, hyper-saline lakes like the Great Salt Lake of the United States do not support fish. Some species of fish have been specially bred to be kept and displayed in an aquarium, and can survive in the home environment.

Hagfish, while generally classified in Agnatha ("jawless") and as fish, actually lack vertebrae. For this reason, hagfish, which are also commonly known as "slime eels," are sometimes not considered to be fish. The other living member of Agnatha, the lamprey, has primitive vertebrae made of cartilage. Hagfish are a staple food in Korea. They are classified in the order Myxini and the family Myxinidae. Both hagfish and lamprey have slimy skin without scales or plates. They also have a notochord that remains throughout life; circular, jawless mouths; and unpaired fins. Hagfish are found in the oceans and lampreys are found in both freshwater and ocean environments. Most lampreys are parasitic.

Fish belonging to the class Chondrichthyes are distinguished by cartilage skeletons, as opposed to skeletons of bone. They have movable jaws and paired fins. Almost all of the Chondrichthyes—sharks, rays, skates, and chimaeras—are found in ocean environments.

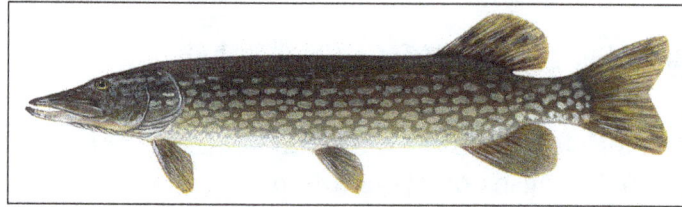

Northern pike, a ray-finned fish.

Most fish species (about 95 percent) are placed in the class Osteichthyes (which some taxonomies consider a superclass). They have bony internal skeletons and skins with scales. (As a general rule for the taxon. Not all bony fish have scales, and scales may be absent or present in two closely related species Catfish is an example of an order of fish that lack scales.) The Osteichthyes taxons include coelacanths (lobe-finned fish), lungfish, and ray-finned fish. Coelacanths were thought to have been extinct for millions of years until fishermen caught one in 1938. Lungfish have lungs, as well as gills. Ray-finned fish are what many people refer to when they use the term fish, as these are our most familiar fish, including bass, eels, and many sports fish. Ray-finned fish have fins that are supported by bones (rays).

Many types of aquatic animals named "fish," such as jellyfish, starfish, and cuttlefish, are not true fish. A number of sea dwelling creatures, like dolphins and whales, are actually mammals.

While fish are poikilothermic in that they do not maintain constant internal temperatures and the temperature often mirrors the ambient temperature, certain species of fish maintain elevated body temperatures to varying degrees. These include teleosts (bony fishes) in the suborder Scombroidei and billfishes, tunas, and one species of "primitive" mackerel (Gasterochisma melampus). All sharks in the family Lamnidae—shortfin mako, long fin mako, white, porbeagle, and salmon shark—are known to have this capacity, and evidence suggests the trait exists in family Alopiidae (thresher sharks). The degree of being able to have elevated temperatures varies from the billfish, which warm only their eyes and brain, to bluefin tuna and porbeagle sharks, which can elevate body temperatures in excess of 20 °C above ambient water temperatures. In many cases, this phenomenon has been traced to heat exchange, as warmer blood being returned to the gills in small veins runs close to colder, oxygenated blood in narrow arteries leaving the gills. This ability to have elevated temperatures allows fish to be active in colder waters and to have enhanced swimming ability because of the warmer muscles. In general, most fish can survive only at a relatively small range of body temperatures, but may adjust their depth in large bodies of water in order to find preferable ranges.

Classification

Fish are a very diverse assemblage, so much so that the term fish is itself more one of convenience than a taxonomic rank. It is used to designate a paraphyletic group, whereby the most recent common ancestor is included but not all descendants, with tetrapods (four-legged vertebrates, or descendants of four-legged vertebrates) being excluded. It is thus not surprising that different taxonomists may classify fish differently.

Latimeria chalumnae (coelacanth, a lobe-finned fish).

Vertebrates are generally classified into two groups, the Agnatha (jawless vertebrates) and the Gnathostomata (jawed vertebrates). The later group includes fish with hinged jaws, but also includes amphibians, reptiles, birds, and mammals (the tetrapods). In most taxonomy, Agnatha and Gnathostomata are each considered a superclass, although sometimes Agnatha is considered a class.

The Agnatha, in addition to including the modern day lampreys (Petromyzontiformes) and hagfish (Myxiniformes), also includes several extinct orders.

Within Gnathostomata, several classes of fish are recognized. Two of these classes have living representatives, the Chondrichthyes (cartilaginous fish) and Osteichthyes (bony fish). In some taxonomy, Osteichthyes is considered a superclass.

Within the Osteichthyes, two extant subclasses (or classes) are generally recognized, the Sarcopterygii (lobe-finned fish) and the Actinopterygii (ray-finned or spiny-finned fish). The coelacanths are generally placed within the Sacropterygii subclass. The Actinopterygii are generally divided into the Chondrostei and the Neopterygii, the latter of which includes the Teleostei (modern bony fishes), a classification into which most fish fit today.

A general grouping of fish, without reference to the names of ranks of taxa (superclass, class, subclass, etc.) is presented above, in the image box. Below is presented a more detailed taxonomic scheme with the rank names, as derived from that offered by Benton, in his text Vertebrate Paleontology:

Subphylum Vertebrata

- Class 'Agnatha'

    ◦ Subclass Myxinoidea (hagfish)

    ◦ Subclass Petromyzontida (lampreys)

    ◦ Subclass Conodonta

    ◦ Subclass Pteraspidomorphi

    ◦ Order Thelodonti

    ◦ Order Anaspida

    ◦ Subclass unnamed

- Order Galeaspida
- Order Pituriaspida
- Order Osteostraci
- Infraphylum Gnathostomata (jawed vertebrates)
  - Class Placodermi
  - Class Chondrichthyes (cartilaginous fish)
    - Subclass Elasmobrachii (sharks, rays, skates)
    - Subclass Subterbranchialia
      - Superorder Holocephali (chimaeras)
  - Class Acanthodii
  - Class Osteichthyes (bony fish)
    - Subclass Actinopterygii (ray-finned fish)
      - Superdivision Chondrostei (sturgeons, paddlefish)
      - Superdivision Neopterygii (teleosts—modern bony fish)
    - Subclass Sarcopterygii (lobe-finned fish)
      - Order Dipnoi (lungfish)
      - Infraclass Crossopterygii
    - Order Actinistia (coelacanths)

Each year, biologists find 200 to 300 species of fish that had not been previously known.

Fish and Humans

Throughout history, humans have utilized fish as a food source. Historically and today, most fish protein has come by means of catching wild fish. However, aquaculture, or fish farming, which has been practiced since about 3,500 B.C.E. in China, is becoming increasingly important in many nations. Overall, about one-sixth of the world's protein is estimated to be provided by fish. That proportion is considerably elevated in some developing nations and regions heavily dependent on the sea. In a similar manner, fish have been tied to trade. One of the world's longest lasting trade histories is the trade of dry cod from the Lofoten area in northern Norway to the southern parts of Europe. This trade in cod has been going on for more than 1000 years.

Fish are also caught for sport. Indeed, in many aquatic environments today, including most fresh-waters, there are more fish caught for sport than via commercial fisheries.

Catching fish for the purpose of food or sport is known as fishing, while the organized effort by humans to catch fish is called a fishery. Fisheries are a huge global business and provide income for millions of people. The annual yield from all fisheries worldwide is about 100 million tons, with popular species including herring, cod, anchovy, tuna, flounder, and salmon. However, the term fishery is broadly applied, and includes more organisms than just fish, such as mollusks and crustaceans, which are often called "fish" when used as food.

Fish have been recognized as a source of beauty for almost as long as used for food, appearing in cave art, being raised as ornamental fish in ponds, and displayed in aquariums in homes, offices, or public settings. As of 2006, there were an estimated 60 million aquarium enthusiasts worldwide.

Basking shark, the second largest living fish (after the whale shark) is a filter feeder that eats zooplankton.

Because of the popularity of fish for food, sport, and hobby, overfishing is a threat to many species of fish. In the May 15, 2004 issue of the journal Nature, it was reported that all large oceanic fish species worldwide had been so systematically overcaught that fewer than 10 percent of 1950 levels remained. Particularly imperiled were sharks, Atlantic cod, Bluefin tuna, and Pacific sardines.

Some fish pose dangers to humans. Although the sharks may be among the most feared, there are actually few shark species that are known to attack humans. The largest sharks, the whale shark and basking shark, are actually plankton feeders. The International Shark Attack File reports there are only about 10-15 deaths each year worldwide. This compares to about 1,000 deaths annually from crocodiles and 60,000 from snakebites.

On the other hand, Smith and Wheeler suggest that, in contrast to prior estimates of 200 venomous fishes, 1,200 species of fish should be presumed venomous. Most of these venomous fishes come from off the coast of eastern and southern Africa, Australia, Indonesia, Phillipines, Polynesia, and southern Japan. About 50,000 people a year suffer from fish stings or envenomations. Perhaps the most dangerous venomous fish is the stonefish, which can release a venomous toxin from spikes on its back when it is provoked or frightened. This toxin can be fatal to humans if not treated promptly. The pufferfish, often better known by the Japanese name Fugu, poses risks to humans because this species contains a highly toxic poison in the internal organs. Despite this, it is considered a delicacy in Japan. The pufferfish needs to be very specially prepared to be safe for eating. Every year a number of people die from consuming this fish.

Barracudas, sea bass, moray eels, and stingrays are among other fish that pose risks to humans in the aquatic environment.

Ichthyology

Ichthyology is the branch of zoology devoted to the study of fish. This includes bony fish (class Osteichthyes, with over 26,000 species), cartilaginous fish (class Chondrichthyes, about 800 species including sharks and rays), and jawless fish (class or superclass Agnatha, about 75 species including lampreys and hagfish).

The study of fish, which is centuries old, reveals humanity's strong and lasting curiosity about nature, with fish providing both inner joy (beauty, recreation, wonder, and religious symbolism) and practical values (ecology, food, and commerce).

With about 27,000 known living species, fish are the most diverse group of vertebrates, with more than one-half of the total vertebrate species. While a majority of species have probably been discovered and described, approximately 250 new species are officially described by science each year.

Hagfish, while generally classified in Agnatha and as fish, actually lack vertebrae, and for this reason sometimes are not considered to be fish. Nonetheless, they remain a focus of ichthyology. Many types of aquatic animals named "fish," such as jellyfish, starfish, and cuttlefish, are not true fish. They, and marine mammals like whales, dolphins, and pinnipeds (seals and walruses) are not a focus of ichthyology.

The practice of ichthyology is associated with aquatic biology, limnology, oceanography, and aquaculture.

Chapter 2

Classification of Fish

Fish are generally classified into three extant classes. These are Agnathan, which are jawless fish, Osteichthyes and Chondrichthyes. Osteichthyes are further subdivided into Sarcopterygii and Actinopterygii. The topics elaborated in this chapter will help in gaining a better perspective about these classes of fish.

Fish, the member of the Animalia Kingdom is classified into Phylum Chordata and Vertebrata Subphylum. Fishes poses notochord, tubular nerve chord, and paired gills, segmentation of the body parts, post anal tail, ventral heart, and an endoskeleton to be the member of the Chordata. In order to be a vertebrate, it poses backbone. This back bone supports and protects the spinal cord.

All the species of the fish found in the world are classified into the following three groups. They are:

- Agnatha - jawless fish
- Chrondrichthyes - cartilaginous fish
- Osteichthyes - bony fish
    ◦ Ray finned group
    ◦ Lobe finned group

About 50 species of Agnatha fish, 600 species of Chrondrichthyes fish and 30,000 species of Osteichthyes fish are found in the world. Most of the fishes in the bony group belong to the ray finned group. According to the biologist there are about 70 fish orders are found in the world.

Sharks and rays; sturgeon and gars; herring-like fishes; trout and salmon; eels, minnows, suckers, and catfish; flying fish and relatives; cod-like fish; flatfish; seahorses and relatives; mullets, silversides, and barracuda; and mackerels and tunas are the main group of fishes.

Agnathan

Agnathan (superclass Agnatha) is any member of the group of primitive jawless fishes that includes the lampreys (order Petromyzoniformes), hagfishes (order Myxiniformes), and several extinct groups.

Hagfishes are minor pests of commercial food fisheries of the North Atlantic, but lampreys, because of their parasitic habit, have been a serious pest of food fisheries in the Great Lakes in North America, where they have reduced the numbers of lake trout and other species. Agnathans are otherwise of little economic importance. The group is of great evolutionary interest, however, because it includes the oldest known craniate fossils and because the living agnathans have many primitive characteristics.

Lamprey

Lampreys (sometimes inaccurately called lamprey eels) are an ancient extant lineage of jawless fish of the order Petromyzontiformes, placed in the superclass Cyclostomata. The adult lamprey may be characterized by a toothed, funnel-like sucking mouth.

Mouth of a sea lamprey, Petromyzon marinus.

There are about 38 known extant species of lampreys and five known extinct species. Parasitic carnivorous species are the most well-known, and feed by boring into the flesh of other fish to suck their blood; but only 18 species of lampreys engage in this micropredatory lifestyle. Of the 18 carnivorous species, nine migrate from saltwater to freshwater to breed (some of them also have freshwater populations), and nine live exclusively in freshwater. All non-carnivorous forms are freshwater species. Adults of the non-carnivorous species do not feed; they live off reserves acquired as ammocoetes (larvae), which they obtain through filter feeding.

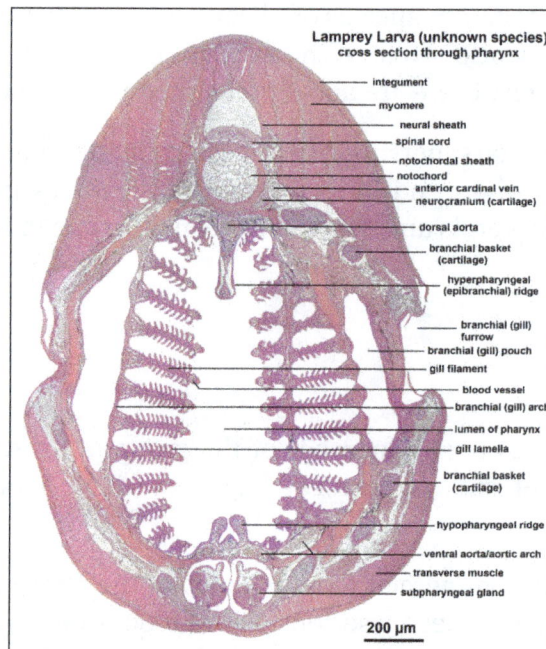

Microscopic cross section through the pharynx of a larva from an unknown lamprey species.

Distribution

Sea lamprey, Petromyzon marinus.

Lampreys live mostly in coastal and fresh waters and are found in most temperate regions except those in Africa. Some species (e.g. *Geotria australis*, *Petromyzon marinus*, and *Entosphenus tridentatus*) travel significant distances in the open ocean, as evidenced by their lack of reproductive isolation between populations. Other species are found in land-locked lakes. Their larvae (ammocoetes) have a low tolerance for high water temperatures, which may explain why they are not distributed in the tropics.

Lamprey distribution may be adversely affected by overfishing and pollution. In Britain, at the time of the Conquest, lampreys were found as far upstream in the River Thames as Petersham. The reduction of pollution in the Thames and River Wear has led to recent sightings in London and Chester-le-Street.

Distribution of lampreys may also be adversely affected by dams and other construction projects due to disruption of migration routes and obstruction of access to spawning grounds. Conversely, the construction of artificial channels has exposed new habitats for colonisation, notably in North America where sea lampreys have become a significant introduced pest in the Great Lakes. Active control programs to control lampreys are undergoing modifications due to concerns of drinking water quality in some areas.

Biology

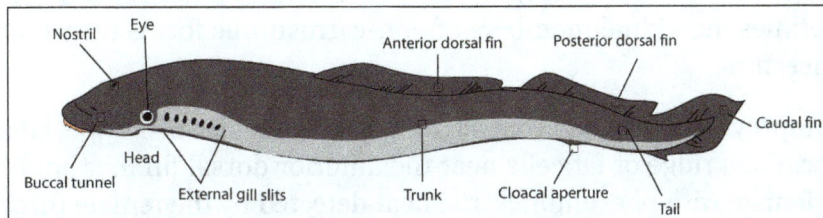

Basic external anatomy of a lamprey.

Adults superficially resemble eels in that they have scaleless, elongated bodies, and can range from 13 to 100 cm (5 to 40 inches) in length. Lacking paired fins, adult lampreys have large eyes, one nostril on the top of the head, and seven gill pores on each side of the head.

The pharynx is subdivided; the ventral part forming a respiratory tube that is isolated from the mouth by a valve called the velum. This is an adaptation to how the adults feed, by preventing the prey's body fluids from escaping through the gills or interfering with gas exchange, which takes place by pumping water in and out of the gill pouches instead of taking it in through the mouth.

Near the gills are the eyes, which are poorly developed and buried under skin in the larvae. The eyes consummate their development during metamorphosis, and are covered by a thin and transparent layer of skin that becomes opaque in preservatives.

The unique morphological characteristics of lampreys, such as their cartilaginous skeleton, suggest they are the sister taxon of all living jawed vertebrates (gnathostomes), and are usually considered the most basal group of the Vertebrata. Instead of true vertebrae, they have a series of cartilaginous structures called arcualia arranged above the notochord. Hagfish, which resemble lampreys, have traditionally been considered the sister taxon of the true vertebrates (lampreys and gnathostomes) but DNA evidence suggests that they are in fact the sister taxon of lampreys.

Studies have shown that lampreys are amongst the most energy-efficient swimmers. Their swimming movements generate low-pressure zones around their body, which pull rather than push their bodies through the water.

The last common ancestor of lampreys appears to have been specialized to feed on the blood and body fluids of other fish after metamorphosis. They attach their mouthparts to the target animal's body, then use three horny plates (laminae) on the tip of their piston-like tongue, one transversely and two longitudinally placed, to scrape through surface tissues until they reach body fluids. The teeth on their oral disc are primarily used to help the animal attach itself to its prey. Made of keratin and other proteins, lamprey teeth have a hollow core to give room for replacement teeth growing under the old ones. Some of the original blood-feeding forms have evolved into species that feed on both blood and flesh, and some who have become specialized to eat flesh and may even invade the internal organs of the host. Tissue feeders can also involve the teeth on the oral disc in the excision of tissue. As a result, the flesh-feeders have smaller buccal glands as they don't require producing anticoagulant continuously and mechanisms for preventing solid material entering the branchial pouches, which could otherwise potentially clog the gills. A study of the stomach content of some lampreys has shown the remains of intestines, fins and vertebrae from their prey. Although attacks on humans do occur, they will generally not attack humans unless starved.

Carnivorous forms have given rise to the non-carnivorous species, and "giant" individuals amongst the otherwise small American brook lamprey have occasionally been observed, leading to the hypothesis that sometimes individual members of non-carnivorous forms return to the carnivorous lifestyle of their ancestors.

Research on sea lampreys has revealed that sexually mature males use a specialized heat-producing tissue in the form of a ridge of fat cells near the anterior dorsal fin to stimulate females. After having attracted a female with pheromones, the heat detected by the female through body contact will encourage spawning.

Lampreys provide valuable insight into adaptive immune systems, as they possess a convergently evolved adaptive immunity with cells that function like the T cells and B cells seen in higher vertebrates. Lamprey leukocytes express surface variable lymphocyte receptors (VLRs) generated from somatic recombination of leucine-rich repeat gene segments in a recombination activating gene-independent manner.

Northern lampreys (Petromyzontidae) have the highest number of chromosomes (164–174) among vertebrates.

Pouched lamprey (*Geotria australis*) larvae also have a very high tolerance for free iron in their bodies, and have well-developed biochemical systems for detoxification of the large quantities of these metal ions.

Lampreys are the only extant vertebrate to have four eyes. Most lampreys have two additional parietal eyes: a pineal and parapineal one (the exception is members of *Mordacia*).

Lifecycle

The adults spawn in nests of sand, gravel and pebbles in clear streams, and after hatching from the eggs, young larvae—called ammocoetes—will drift downstream with the current till they reach soft and fine sediment in silt beds, where they will burrow in silt, mud and detritus, taking up an existence as filter feeders, collecting detritus, algae, and microorganisms. The eyes of the larvae are underdeveloped, but are capable of discriminating changes in illuminance. Ammocoetes can grow from 3–4 inches (8–10 cm) to about 8 inches (20 cm). Many species change color during a diurnal cycle, becoming dark at day and pale at night. The skin also has photoreceptors, light sensitive cells, most of them concentrated in the tail, which helps them to stay buried. Lampreys may spend up to eight years as ammocoetes, while species such as the Arctic lamprey may only spend one to two years as larvae, prior to undergoing a metamorphosis which generally lasts 3–4 months, but can vary between species. While metamorphosing, they do not eat.

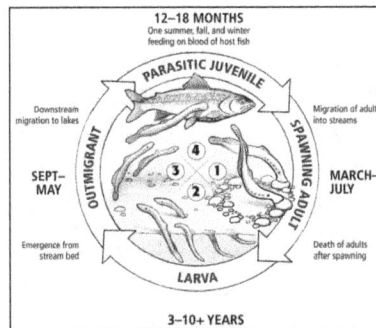

Sea lamprey

The rate of water moving across the ammocoetes' feeding apparatus is the lowest recorded in any suspension feeding animal, and they therefore require water rich in nutrients to fulfill their nutritional needs. While the majority of (invertebrate) suspension feeders thrive in waters containing under 1 mg suspended organic solids per litre (>1 mg/l), ammocoetes demand minimum 4 mg/l, with concentrations in their habitats having been measured up to 40 mg/l.

During metamorphosis the lamprey loses both the gallbladder and the biliary tract, and the endostyle turns into a thyroid gland.

Some species, including those that are not carnivorous and do not feed even following metamorphosing, live in freshwater for their entire lifecycle, spawning and dying shortly after metamorphosing. In contrast, many species are anadromous and migrate to the sea, beginning to prey on other animals while still swimming downstream after their metamorphosis provides them with eyes, teeth, and a sucking mouth. Those that are anadromous are carnivorous, feeding on fishes or marine mammals.

Anadromous lampreys spend up to four years in the sea before migrating back to freshwater, where they spawn. Adults create nests (called redds) by moving rocks, and females release thousands of eggs, sometimes up to 100,000. The male, intertwined with the female, fertilizes the eggs simultaneously. Being semelparous, both adults die after the eggs are fertilized.

Hagfish

Hagfish is the common name for the marine craniates (animals with skulls) of the class (or sub-phylum) Myxini, characterized by a scaleless, eel-like body that lacks both paired fins and verte-brae, but does have a cranium. Hagfish are the only animals that have a skull but not a vertebral column. Despite their lack of vertebrae, the hagfish have traditionally been classified among the vertebrates. This traditional classification is currently under dispute.

Although hagfish have an ancient history, possibly tracing back 300 million years ago, to the Car-boniferous, there remain today living species of hagfish. These animals, which are characterized by degenerate eyes, barbels present around the mouth, and teeth only on the tongue, are found in marine environments and are scavengers that eat primarily the insides of dying or dead fish and invertebrates. They are unique in being the only "vertebrate" in which the body fluids are isosmotic (having the same osmotic pressure) with seawater. Although hagfish are sometimes called "slime eels," they are not eels at all, which are part of the bony fish.

The unusual feeding habits and slime-producing capabilities of hagfish have led members of the scientific and popular media to dub the hagfish as the most "disgusting" of all sea creatures. None-theless, they play an important ecological role and for humans are of value for both commerce and research. In food chains, hagfish are consumed by seabirds, pinnipeds (seals and walruses), and crustaceans such as lobsters and crabs. In some areas of the world, they are consumed by people, being important commercially in Korea for this reason. They are used also in the study of tumors and for investigating chordate relationships through genetic analyses. Scientists are exploring pos-sible practical applications for the hagfish slime. Finally, their mysterious and unusual behaviors and forms, and their connection to an ancient past, add greatly to the wonder of nature.

A group of Pacific hagfish.

The hagfish family, Myxinidae, is the only family in the order Myxiniformes (also known as Hyper-otreti), which itself is the only order in the class Myxini. Thus, hagfish is variously used for any of the three taxonomic levels.

Hagfish are jawless and generally classified with the lampreys into the superclass Agnatha (jawless vertebrates) within the subphylum Vertebrata. However, hagfish actually lack vertebrae. For this reason, they sometimes are separated from the vertebrates and not even considered to be fish. Jan-vier and a number of others put hagfish in a separate subphylum Myxini, which along with the sub-phylum Vertebrata comprises the taxon Craniata, recognizing the common possession of a crani-um. Others, however, place Vertebrata and Craniata as synonyms at the same level of classification, and thereby retain hagfish (Myxini) as members of the superclass Agnatha within the vertebrates. The other living member of Agnatha, the lamprey, has primitive vertebrae made of cartilage.

Yet other classifications place Myxini as a class that in one instance lies within the subphylum Vertebrata and, in another instance, lies within—and is the only class in—the clade Craniata, which is considered to be separate from the subphylum Vertebrata.

As members of Agnatha hagfish are characterized by the absence of jaws derived from gill arches (bone or cartilage supporting the gills), although they do have a biting apparatus that is not considered to have been derived from gill arches. Other common characteristics of Agnatha include the absence of paired fins, absence of pelvic fins, the presence of a notochord both in larvae and adults, and seven or more paired gill pouches. In addition, the gills open to the surface through pores rather than slits, and the gill arch skeleton is fused with neurocranium (the portion of the skull the protects the brain).

Despite their name, there is some debate about whether hagfish are strictly fish, since they belong to a much more primitive lineage than any other group that is commonly defined as fish (Chondrichthyes and Osteichthyes), and because of the lack of a vertebrae commonly associated with the definition of fish.

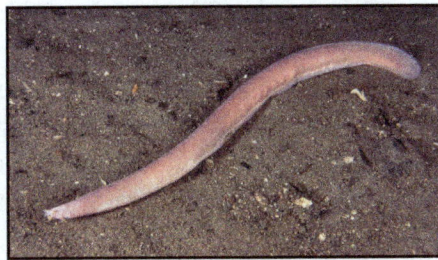
Drawing of a New Zealand hagfish.

Extant hagfish are placed in the family Myxinidae within the order Myxiniformes (Hyperotreti) and subphylum or class Myxini.

Members of the order Myxiniformes are characterized by one semicircular canal, absence of eye musculature, a single olfactory capsule with few folds in the sensory epithelium, no bone, and one to 16 pairs of external gill openings.

Members of the family Myxinidae are characterized by the lack of a dorsal fin, barbels present around the mouth, degenerate eyes, teeth only on the tongue, no metamorphosis, and ovaries and testes in the same individual but only one functional gonad. Note that many of these features differ from the other agnathan, lampreys, which have one or two dorsal fins, well-developed eyes, absence of barbels, separate sexes, a larval stage that undergoes radical metamorphosis, and teeth both on the oral disc and tongue.

Hagfish have a scaleless, elongated, cel-like body without paired fins. The extant hagfish average about half a meter (18 inches) long. The largest known species is Eptatretus goliath with a specimen recorded at 127 centimeters, while Myxine kuoi and Myxine pequenoi seem to reach no more than 18 centimeters.

Extant hagfish have paddle-like tails, cartilaginous skulls, and tooth-like structures composed of keratin. Colors range from pink to blue-gray depending on the species, and may include black or white spots. Eyes may be vestigial or absent. Hagfish have no true fins and have six barbels around the mouth and a single nostril. Instead of vertically articulating jaws like Gnathostomata (vertebrates with jaws), they have a pair of horizontally moving structures with tooth-like projections for pulling off food.

The circulatory systems of the extant hagfish have both closed and open blood vessels, with a heart system that is more primitive than that of vertebrates, bearing some resemblance to that of some worms. This system comprises a "brachial heart," which functions as the main pump, and three types of accessory hearts: The "portal" heart(s), which carry blood from intestines to liver; the "cardinal" heart(s), which move blood from the head to the body; and the "caudal" heart(s), which pump blood from the trunk and kidneys to the body. None of these hearts are innervated, so their function is probably modulated, if at all, by hormones.

Slime

Extant hagfish are long, vermiform (worm-like), and can exude copious quantities of a sticky slime or mucus (from which the typical species Myxine glutinosa was named). There are from 70 to 200 slime glands found in each of two ventrolateral lines from head to tail. The slime glands contain both mucous cells and thread cells, with the thread from the thread cells probably adding tensile strength to the slime. Indeed, the mucus excreted by hagfish is unique in that it includes strong, threadlike fibers similar to spider silk, causing the slime to be fiber-reinforced. No other slime secretion known is reinforced with fibers in the same manner as hagfish slime. The fibers are about as fine as spider silk (averaging two micrometers), but can be 12 centimeters long. When the coiled fibers leave the gland, they unravel quickly to their full length without tangling.

When captured and held by the tail, hagfish escape by secreting the fibrous slime, which turns into a thick and sticky gel when combined with water. They clean off the slime by tying themselves in an overhand knot, which works its way from the head to the tail of the animal, scraping off the slime as it goes. Some authorities conjecture that this singular knotting behavior may assist them in extricating themselves from the jaws of predatory fish. The "sliming" also seems to act as a distraction to predators, and free-swimming hagfish are seen to "slime" when agitated and will later clear the mucus off by way of the same traveling-knot behavior.

The slime appears to be particularly effective at clogging gills of fish and thus it is speculated that sliming may be an effective defense mechanism against fish, which are not among the main hagfish predators.

An adult hagfish can secrete enough slime to turn a large bucket of water into gel in a matter of minutes.

Behavior and Reproduction

Hagfish tend to burrow under rocks or into mud, in highly saline waters, and away from bright light. They are mostly found either near the mouths of rivers or at depths of 25 or more meters, with Myxiine circifrons found at more than 1000 meters below the ocean surface.

Hagfish typically are scavengers, entering decaying and dead fish and invertebrates (including (polychaete marine worms and shrimp), feeding on the insides. Living organisms also are consumed. While having no ability to enter through skin, they often enter through natural openings such as the mouth, gills, or anus and consume their prey from the inside out. They can be a great nuisance to fishermen, as they are known to infiltrate and devour a catch before it can be pulled to the surface.

Like leeches, they have a sluggish metabolism and can survive months between feedings.

Very little is known about hagfish reproduction. In some species, the sex ratio can be as high as 100 to 1 in favor of females. In other species, individual hagfish that are hermaphroditic are not uncommon. These individuals have both ovaries and testes, but the female gonads remain non-functional until the individual has reached a particular stage in the hagfish lifecycle. Females typically lay 20 to 30 yolky eggs that tend to aggregate due to having Velcro-like tufts at either end.

Hagfish do not have a larval stage, in contrast to lampreys, which have a long larval phase.

Chondrichthyes

Bull shark (Carcharhinus leucas).

Chondrichthyan, (class Chondrichthyes), also called chondrichthian is any member of the diverse group of cartilaginous fishes that includes the sharks, skates, rays, and chimaeras. The class is one of the two great groups of living fishes, the other being the osteichthians, or bony fishes. The name Selachii is also sometimes used for the group containing the sharks.

Remora; shark: A remora (Echeneis naucrates) and its host, a zebra shark (Stegostoma fasciatum). By attaching itself to the shark, the remora is carried along by the shark, allowing the remora to travel to different areas without having to expend its own energy to swim. The shark is completely unaffected by the remora's presence.

Many unique structural, physiological, biochemical, and behavioral characters make these fishes of particular interest to scientists. These fishes are, in a sense, living fossils, for many of the living sharks and rays are assigned to the same genera as species that swam the Cretaceous seas over 100 million years ago. More than 400 species of sharks and about 500 species of rays are known. Although by any reckoning a successful group, the modern chondrichthyans number far fewer species than the more advanced bony fishes, or teleosts.

The danger some sharks and stingrays present to humans makes these animals fascinating and, at the same time, fearsome. Perhaps for this reason, they figure prominently in the folklore and art of many tropical peoples who depend on the sea. The danger from shark attack, while very real, is remarkably uncommon and easily sensationalized. Quite frequently, little attempt is made to distinguish between dangerous and harmless species.

General Features

Problems of Taxonomy

The name Selachii refers to a category of fishlike vertebrates characterized by a skeleton primarily composed of cartilage. Selachii are given a variety of treatments by ichthyologists. Some authorities consider Selachii to be a class or subclass that contains all the modern sharks and rays; other authorities restrict the use of the name to an order of modern sharks and certain extinct ancestral forms. Under the latter system, the rays (including the sawfishes, guitarfishes, electric rays, mantas, skates, and stingrays) are ranked separately.

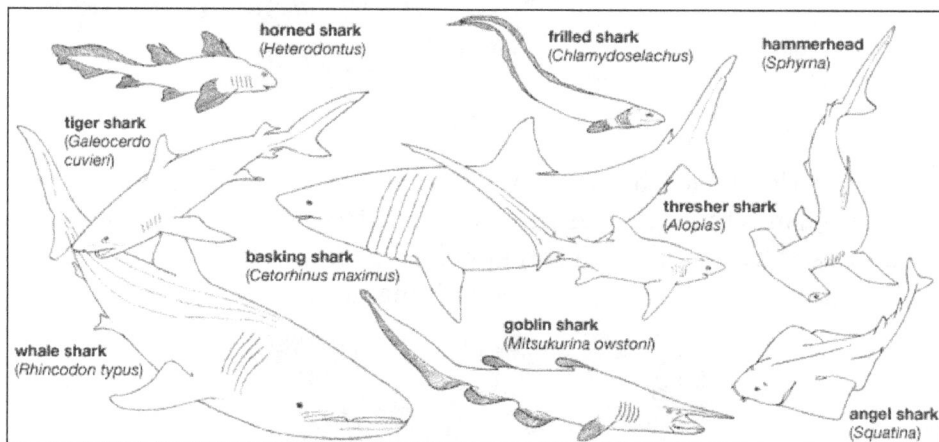

Body plans of representative sharks.

The chimaeras (Holocephali) bear many similarities to sharks and rays in skeletal structure, internal organs, and physiology. Ichthyologists commonly although not unanimously emphasize these similarities by grouping the modern and ancient sharks, rays, and chimaeras in the class Chondrichthyes, the cartilaginous fishes. Under this system, the sharks, skates, and rays are further grouped into one subclass, Elasmobranchii, and the chimaeras into another, Holocephali. Some authorities classify the elasmobranchs into one class (Selachii) and classify the chimaeras into another (Holocephali); however, assigning the two groups class rank implies a degree of distinctness equal to that of the amphibians (Amphibia), reptiles (Reptilia), birds (Aves), and mammals (Mammalia).

Distribution and Abundance

The majority of sharks and rays are marine fishes, but many enter estuaries; some travel far up rivers, and a few are permanent residents of fresh water. Most species live in the relatively shallow waters of continental margins or around offshore islands; a few roam far out in the vast spaces of the oceans. Some live at great depths, in midwaters or on the bottom; others are surface swimmers or inhabit the bottom in shallow waters.

Once regarded as "trash" fish, sharks and rays are increasingly represented in the fisheries of most countries. With numbers of more highly valued bony fishes decreasingly rapidly, many fisheries are specifically targeting elasmobranchs as a primary commercial resource. Annual yields are as much as 750,000 metric tons (roughly 827,000 short tons), and most of this is sold fresh, dried, salted, or processed. This intense harvest is causing the reduction of many shark and ray populations and, in some cases, causing the commercial extinction of some species. One consequence of this depletion of these top-level predators may be a disruption of the food chain in marine ecosystems.

Food Habits

Sharks

All sharks are carnivorous and, with a few exceptions, have broad feeding preferences, governed largely by the size and availability of the prey. The recorded food of the tiger shark (Galeocerdo cuvier), for example, includes a wide variety of fishes (including other sharks, skates, and sting-rays), sea turtles, birds, sea lions, crustaceans, squid, and even carrion such as dead dogs and garbage thrown from ships. Sleeper sharks (Somniosus), which occur mainly in polar and subpolar regions, are known to feed on fishes, small whales, squid, crabs, seals, and carrion from whaling stations. Many bottom-dwelling sharks, such as the smooth dogfishes (Triakis and Mustelus), take crabs, lobsters, and other crustaceans, as well as small fishes.

The three largest sharks, the whale shark (Rhincodon typus), the basking shark (Cetorhinus maximus), and the megamouth shark (Megachasma pelagios), resemble the baleen whales in feeding mode as well as in size. They feed exclusively or chiefly on minute passively drifting organisms (plankton). To remove these from the water and concentrate them, each of these species is equipped with a special straining apparatus analogous to baleen in whales. The basking shark and the megamouth shark have modified gill rakers, the whale shark elaborate spongy tissue supported by the gill arches. The whale shark also eats small, schooling fishes.

The saw sharks (Pristiophoridae) and sawfishes (Pristidae), though unrelated, both share a specialized mode of feeding that depends on the use of their long bladelike snout, or "saw." Equipped with sharp teeth on its sides, the saw is slashed from side to side, impaling, stunning, or cutting the prey fish. Saw sharks and sawfishes, like most other rays, are bottom inhabitants.

Thresher sharks (Alopias) feed on open-water schooling fishes, such as mackerel, herring, and bonito, and on squid. The long upper lobe of the tail, which may be half the total length of the shark, is used to herd the fish (sometimes by flailing the water surface) into a concentrated mass convenient for feeding. Thresher sharks have also been observed to stun larger fish with a rapid strike of the tail.

Most sharks and rays do not school. Individuals are normally solitary and usually come together only to exploit food resources or to mate. During these encounters, some species may show specific dominance structures, usually based on size. Some species, however, will travel in large schools segregated by size, a habit that protects smaller individuals from being eaten by larger ones. Still other species form sex-segregated schools where males and females live in slightly different habitats or depths. When potential prey is discovered, sharks circle it, appearing seemingly out of nowhere and frequently approaching from below. Feeding behaviour is stimulated by increasing numbers and rapid swimming, when three or more sharks appear in the presence of food. Activity soon progresses from tight circling to rapid crisscross passes. Biting habits vary with feeding methods and dentition. Sharks with teeth adapted for shearing and sawing are aided in biting by body motions that include rotation of the whole body, twisting movements of the head, and rapid vibrations of the head. As the shark comes into position, the jaws are protruded, erecting and locking the teeth into position. The bite is extremely powerful; a mako shark (Isurus), when attacking a swordfish too large to be swallowed whole, may remove the prey's tail with one bite. Under strong feeding stimuli, the sharks' excitement may intensify into what is termed a feeding frenzy, possibly the result of stimulatory overload, in which not only the prey but also injured members of the feeding pack are devoured.

In most cases, sharks locate food by smell, which is well developed in nearly all species. Sharks also possess other important senses that allow them to find food, and the importance of each sense varies between species. Their lateral line system, a series of sensory pores along the side of the body for detecting vibrations, allows sharks to detect vibrations in the water. Their network of ampullae allows them to sense weak electrical signals given off by prey, and their eyes are often acute enough to discriminate the size, shape, and colour of their prey. The sum of these senses working together makes a well-integrated system for finding prey.

Rays

The majority of batoid fishes (members of the order Batoidei such as rays and allies) are bottom dwellers, preying on other animals on or near the seafloor. Guitarfishes (Rhynchobatidae and Rhinobatidae), butterfly rays (Gymnuridae), eagle rays (Mylobatidae), and cow-nosed rays (Rhinopteridae) feed on invertebrates, principally mollusks and crustaceans. Whip-tailed rays (Dasyatidae) use their broad pectoral fins to dig shellfish from sand or mud. Skates (Rajidae) lie on the bottom, often partially buried, and rise in pursuit of such active prey as herring. Skates trap their victims by swimming over and then settling upon them, a practice facilitated by their habit of hunting at night.

Electric rays (Torpedinidae) are characteristically bottom fishes of sluggish habits. They feed on invertebrates and fish, which may be stunned by shocks produced from the formidable electric organs. With their electricity and widely extensible jaws, these rays are capable of taking very active fishes, such as flounder, eel, salmon, and dogfish. Shallow-water electric rays have been observed to trap fishes by suddenly raising the front of the body disk while keeping the margins down, thereby forming a cavity into which the prey is drawn by the powerful inrush of water.

Most of the myliobatoid rays (seven recognized families of the suborder Myliobatoidei [order Myliobatiformes], which includes all the typical rays) swim gracefully, with undulations of the broad winglike pectoral fins. Some species, especially the eagle rays, frequently swim near the surface and even jump clear of the water, skimming a short distance through the air.

Manta, or devil, rays (Mobulidae) swim mostly at or near the surface, progressing by flapping motions of the pectoral fins. Even the largest often leap clear of the water. In feeding, a manta moves through masses of macroplankton or schools of small fish, turning slowly from side to side and using the prominent cephalic fins, which project forward on each side of the mouth, to funnel the prey into the broad mouth.

Chimaeras and ghost sharks (Chimaeridae) dwell near the bottom in coastal and deep waters, to depths of at least 2,500 metres (about 8,000 feet). They are active at night, feeding almost exclusively on small invertebrates and fishes.

Reproductive Behavior

Mature individuals of some species of sharks segregate by sex, coming together only during the mating season, when the males—at least those of the larger, more aggressive species—stop feeding. Segregation is a behavioral adaptation to protect the females. One principal courting activity used by the male to induce cooperation of the female in mating is the act of biting her and gripping her with his teeth. A male takes hold of a female in this way so that he can more easily insert a modified fin, called a clasper, into her cloaca. After mating, the sexes again separate. The pregnant females tend to keep apart from the other females of like size. As the time of parturition approaches, the pregnant females move to particular areas, which presumably have environmental properties especially suitable as nursery grounds. When giving birth to their young, they stop feeding, and, soon after parturition is completed, they depart.

Nursery areas vary with species. Some sharks—such as the bull shark (Carcharhinus leucas) and the sandbar shark (C. plumbeus)—use shallow waters of bays and estuaries; the silky shark (C. falciformis) uses the bottom far out on oceanic banks such as the Serrana Bank in the western Caribbean. The Atlantic spiny dogfish (Squalus acanthias) bears its young mostly during the winter, far out on the continental shelf of northeastern America, almost two years after mating.

A few skates that have been observed mating may be characteristic of other rays. The male seizes the female by biting the pectoral fin and presses his ventral surface against hers while inserting his clasper or in some species both claspers, into her cloaca. Male skates have one to five rows of clawlike spines on the dorsal side of each pectoral fin. These are retractile in grooves of the skin and are used to hold the female during mating.

The eggs of skates in aquaria have been observed to be extruded in series, usually of two eggs at a time but sometimes one. Rest periods of one to five days occur between extrusions. A female of a European skate, Raja brachyura, laid 25 eggs over a 49-day period in the National Marine Aquarium, located in Plymouth.

Although the mating of chimaeroids has not been observed, it is generally presumed that the mode of copulation is similar to that of sharks and that the male's frontal spine and anterior appendage of the pelvic fins are probably used in securing the female. Two eggs are laid simultaneously, one from each oviduct. They are often carried for a relatively long period before being laid, several hours or even days, each egg protruding from the female for the greater part of its length.

Form and Function

Distinguishing Features

The elasmobranchs are fishlike vertebrates that differ from bony fishes in many respects. The skeleton is composed of cartilage and, although often calcified (especially in the vertebrae), lacks true bone (except in the roots of teeth). There are five to seven fully developed gill clefts, opening separately to the exterior. Most sharks and all rays have an opening behind each eye, called a spiracle, which is a modified first gill cleft. The dorsal fin or fins and fin spines are rigid, not erectile. Scales, if present, are structurally minute teeth, called dermal denticles, each consisting of a hollow cone of dentine surrounding a pulp cavity and covered externally by a layer of hard enamel-like substances called vitrodentine. The scales covering the skin do not grow throughout life, as they do in bony fishes, but have a limited size; new scales form between existing ones as the body grows. Certain other structures, such as the teeth edging the rostrum (beak) of sawfishes and saw sharks, the stinging spines of stingrays, and the teeth in the mouth, are structurally modified scales. The teeth, arranged in rows in the mouth, are not firmly attached to the jaws but are imbedded in a fibrous membrane lying over the jaws. When a tooth becomes broken, worn, or lost, it is replaced by one moving forward from the next row behind; at the base of the innermost row are rudimentary teeth and tooth buds that develop and move forward as needed. A spiral membranous fold (spiral valve) extends through the intestine of all sharks, rays, and chimaeras.

The rays differ externally from sharks in having the gill openings confined to the lower surface; the eyes of the rays are on the dorsal surface, and the edges of the pectoral fins are attached to the sides of the head in front of the gill openings. Some rays lack scales, and others are variously armed with thorns, tubercles, or prickles, all of which are modified scales; the tails of some have long, saw-toothed spines equipped with poison glands. In the sawfishes the snout is prolonged into a long, flat blade armed on either side with teeth. The electric rays have electric organs by which they can administer electric shocks to enemies or prey.

The chimaeras have only one external gill opening. In the adult the skin on each side of the head is smooth and lacks scales; the teeth consist of six pairs of grinding plates. The dorsal fin and spine are erectile. Like male sharks and rays, male chimaeras have claspers that serve to transfer sperm to the female, but in addition they have an erectile clasping device, the tantaculum, in front of each pelvic fin; most species have another such organ on top of the head.

Senses

Although sharks are often said to have a low order of intelligence, they, as well as rays and chimaeras, have survived successfully over a long period of geologic time. They are well equipped to locate prey and their own kind; to direct the course of their seasonal migrations; to discriminate specific localities; to respond to variations of temperature; to react to attractive or repelling substances in the water; and perhaps even to feel objects some distance away from them. They can see, hear, smell, taste, feel, and maintain their equilibrium. The roles of the sense organs have been studied in only a few species, principally sharks, and consequently remain imperfectly understood.

The sense of smell is highly developed and probably the principal means of locating prey and guiding the predator toward it. Given a favourable direction of current, sharks can detect incredibly minute concentrations—fractions of a part per million (that is, less than 1×10^{-6} parts)—of certain substances in the water, such as blood.

Although their eyes are structurally and functionally adapted for seeing, it is believed that their visual acuity in discerning the form and colour of an object varies between species. The importance of sight is relative to the habitat and feeding habits of each species. Fast predatory sharks tend to have more acute vision, and in some deep-diving species the eyes are well developed to maximize detection of ambient light.

The hearing apparatus, located in the auditory capsule of the cranium, includes a system of semi-circular canals, which are responsible for maintaining equilibrium. Sharks seem to be remarkably sensitive to sounds of low frequency and to possess extraordinary faculty for directional hearing. Whether hearing is more sensitive than smell has not yet been established.

Sensory organs identified as taste buds are located on the floor, sides, and roof of the mouth and on the throat, as well as on the tongue. Experiments on several species of large sharks indicate that they do discriminate food types—preferring tuna, for example, to other fish species. Under some conditions, however, they become less fastidious, going into a feeding frenzy in which they attack anything, including others of their own kind.

Sensory organs located in the skin of all sharks, rays, and chimaeras receive a variety of information—vibrations of low frequencies, temperature, salinity, pressure, and minute electrical stimuli, such as that produced by another fish in the vicinity. These sensory organs are located in the lateral line system, in groups of pores called ampullar organs, found on the head, snout, and around the jaws, which detect electrical impulses.

Salt and Water Balance

Most marine vertebrates maintain lower concentrations of salts and other chemicals in their blood than are found in seawater. As a result, these animals face a continuous problem of water loss to the environment, because of the tendency of water to move through membranes from regions of low salt concentration to regions of higher concentration. The marine cartilaginous fishes differ from almost all of the bony fishes (except the coelacanths and aestivating lungfishes) in being able to reabsorb in the renal (kidney) tubules most of their nitrogenous waste products (urea and trimethylamine oxide) and to accumulate these products in their tissues and blood, an ability termed the urea retention habitus. The concentration within the body thus exceeds that of the surrounding seawater, and water moves into the body with no expenditure of energy. When any of these fishes moves into fresh water, as many do, the urine flow to the outside increases; hence, the concentration of urea in the blood decreases. In the sawfish, for example, the increase of urine output is more than twentyfold; the blood urea concentration decreases to less than one-third the amount observed in marine forms. Purely freshwater elasmobranchs, such as the stingrays of the Orinoco and Amazon drainage systems, seem to lack the urea retention habitus.

Respiration

Sharks breathe chiefly by opening the mouth while expanding the mouth-throat (bucco-pharyngeal) cavity and contracting the gill pouches to close the gill slits. With the mouth closed, they contract the bucco-pharyngeal cavity while dilating the gill pouches, thus drawing the water over the gills where the exchange of oxygen and carbon dioxide takes place. Then, with the mouth still closed, they contract the bucco-pharyngeal cavity and gill pouches, and the gill slits are opened to expel the water.

Most of the rays, on the other hand, take in water chiefly through the spiracles; these then close by contraction at their anterior margins, which bear rudimentary gill filaments and a spiracular valve. Folds of membrane on the roof and floor of the mouth prevent the water from passing down the throat and direct it to the gill openings. Skates, which sometimes hold the lower surface of the head slightly above the bottom, may inhale some water through the mouth; mantas, which have small spiracles and live near the surface, respire chiefly through the mouth. Skates, stingrays, guitarfishes, and angel sharks frequently reverse the direction of flow through the spiracles, apparently to clear them of foreign matter.

Chimaeras take in water chiefly through the nostrils, keeping the mouth closed for the most part. The water reaches the mouth primarily through grooves leading there from the nostrils.

Reproduction and Development

All species of sharks, rays, and chimaeras produce large yolk-rich eggs. These are fertilized internally, for which the males are equipped with two copulatory organs called claspers along the inner edges of the pelvic fins. Each clasper has a groove for guidance of sperm. The few published descriptions of mating sharks and rays are probably characteristic of the entire group. The male grasps one of the female's pectoral fins with his teeth to hold her in position as he inserts a clasper through a cavity (cloaca) and into a tube (oviduct). Males of most species probably use only one clasper at a time. The sperm travel to the anterior end of the oviduct, where they fertilize the eggs. The eggs then move down the oviduct past the shell gland, where they are covered by a shell or capsule.

Some of the sharks, probably all the skates, possibly some of the guitarfishes, and all of the chimaeras are oviparous (egg-laying species). The eggs are enveloped in a horny shell, usually equipped with tendrils for coiling around solid objects or with spikelike projections for anchoring in mud or sand. The egg cases of most species are more or less pillow-shaped; those of the horned sharks (Heterodontus francisci) are screw-shaped with a spiral flange. The eggs of chimaeras are elliptic, spindle-shaped, or tadpole-shaped and open to the exterior through pores and slits that permit entrance of water during incubation. An egg of the whale shark found in the Gulf of Mexico measured 30 cm (12 inches) long by about 14 cm (5.5 inches) wide and was 8 cm (3 inches) thick. Protected by the shell and nourished by the abundant yolk, the embryo of an oviparous species develops for 18 to 59 weeks before hatching.

The majority of sharks and rays other than the skates are ovoviviparous (that is, the egg hatches within the mother). In this case, the egg is first coated in the shell gland with a temporary membranous capsule that lasts only during early development. After emerging from its capsule, the embryo remains in the oviduct of the mother, nourished by the yolk sac to which it remains attached. Embryos of some ovoviviparous sharks, notably the porbeagle (Lamna nasus), the mako (Isurus oxyrinchus), and the sand shark (Odontaspis taurus), ingest yolks of other eggs and even other embryos within the oviduct of the mother after the contents of their own yolk sacs are exhausted. In the majority of ovoviviparous sharks and rays, organically rich uterine secretions provide supplemental nourishment, which is absorbed by the yolk sac and in many cases by appendages borne on its stalk. In some genera of rays, vascular filaments producing these secretions extend through the spiracles and into the digestive tract of the embryos.

Several shark species are viviparous—that is, the yolk sac develops folds and projections that interdigitate with corresponding folds of the uterine wall, thus forming a yolk-sac placenta through which nutrient material is passed from the mother.

Growth

Growth of a few shark species has been measured or estimated by the differences in length at the times of tagging and recapturing specimens. Growth is also measured by the statistical analysis of the length in systematically collected samples, by the space between concentric circles on the centra of the vertebrae, and by periodic measurements of specimens kept in aquariums. All studies indicate a slow growth rate. During the 10 years between birth and maturity, male Atlantic spiny dogfish grow an average of 47 cm (19 inches) and females 67 cm (26 inches). The Greenland shark (Somniosus microcephalus), which attains 6.5 metres (21 feet) or more (although rarely taken larger than about 4 metres [13 feet]), grows only about 7.5 mm (about 0.3 inch) per year. The annual growth increments of tagged juvenile whitetip reef and Galapagos sharks, both species that become at least 2.5 metres (8 feet) long, were found to be 31 to 54 mm (1 to 2 inches) and 41 mm (about 1.5 inches), respectively. The Australian school shark (Galeorhinus australis) grows about 80 mm (3 inches) in its first year and about 30 mm (1 inch) in its 12th year. By its 22nd year, it is estimated to be approaching its maximum length of 1.6 metres (about 5 feet).

The disk of the eastern Pacific round stingray (Urolophus halleri) increases in width on the average from 75 mm (3 inches) at birth to 150 mm (6 inches) when mature (that is, at 2.6 years old). In the next five years it grows about 60 mm (about 2.4 inches) more toward its maximum recorded width of 25 cm (10 inches) in males or 31 cm (12.25 inches) in females. The males of European thornback rays (Raja clavata) are about 50 cm (20 inches) wide when they reach first maturity, about seven years after birth; females are 60 to 70 cm (24 to 28 inches) at first maturity, nine years after birth.

Osteichthyes

Osteichthyes, known as the bony fish, are a taxonomic class (or superclass) of fish and the largest class of vertebrates in existence today. With over 26,000 species, they comprise over 95 percent of all fish species. The Osteichthyes include the ray-finned fish (subclass or class Actinopterygii) and lobe finned fish (subclass or class Sarcopterygii). However, some taxonomic schemes do not consider Osteichthyes a formal taxonomic group, as it is not monophyletic, although the term is retained for vernacular use for those actinopterygians and sarcopterygians conventionally termed fishes.

The members of Osteichthyes represent an extraordinary diversity of forms, including the "living fossil" coelacanths, long "sea-serpent" oarfish, tiny cichlids, whiskered catfish, colorful clownfish, voracious piranhas, unique seahorses, and so forth. This diversity of forms, behaviors, and colors provides aesthetic joy to humans, as well as a variety of practical uses—food, sportfishing, aquarium fish, and so forth.

Although lampreys, sharks, and rays are also fish, some individuals commonly use the term only in reference to the jawed bony fish, that is, members of Osteichthyes.

Atlantic herring.

Classification

Fish are a very diverse assemblage, so much so that the term fish is itself more one of convenience than a taxonomic rank. In general, a fish can be defined as a poikilothermic (cold-blooded), water dwelling vertebrate with gills throughout life, and limbs—if present—in the form of fins. Poikilothermic refers to the fact that the internal temperatures of fish vary, often matching the ambient temperature of the environment. (However, some fish, such as the blue-fin tuna, can flourish at significantly higher internal temperatures than the surrounding water.)

Modern fish are conventionally divided into the jawless fish (class or superclass Agnatha, with about 75 species including lampreys and hagfish), the cartilaginous fish (class Chondrichthyes, with about 800 species including sharks and rays), and the bony fish (class or superclass Osteichthyes, with over 26,000 species). Some individuals do not include as fish those members of Agnatha or Chondrichthyes.

Common Clownfish (Amphiprion ocellaris) in their magnificent
sea anemone (Heteractis magnifica) home.

With over 27,000 known extant (living) species, fish constitute more than one-half of the total vertebrate species. The vast majority of fish are members of Osteichthyes. Nelson recognizes 45 orders, 435 families, 4079 genera, and 23,689 species of osteichthyans (bony fishes).

Within the Osteichthyes, two extant subclasses (or classes) are generally recognized: the Sarcopterygii (lobe-finned fish) and the Actinopterygii (ray-finned or spiny-finned fish). Most bony-fish belong to the Actinopterygii; there are only eight living species of lobe finned fish, including the lungfish (6 extant species) and coelacanths (2 extant species). Some species of lobe-finned fish have jointed bones. Actinopterygii, or ray-finned fish, include the most familiar fish, such as sturgeons, gars, eels, carp, herrings, anchovies, catfishes, goldfishes, piranhas, oarfish, seahorses, bass, cichlids, pickerel, salmon, and trout.

The Actinopterygii are generally classified into two groups—the Chondrostei and the Neopterygii—the latter of which includes the Teleostei (modern bony fishes), a classification into which most fish fit today. The Sarcopterygii are traditionally divided into the lungfish and coelacanths.

While Osteichthyes traditionally is treated as a class of vertebrates, with subclasses Actinopterygii and Sarcopterygii, some newer schemes divide this group into separate classes. Nelson, for example, lists Actinopterygii and Sarcopterygii as classes.

However, while using Osteichthyes as vernacular for actinopterygians and those sarcopterygians conventionally termed fishes, Nelson prefers not to use the term as a formal taxon as it "is clearly not a monophyletic group." Rather, Nelson uses the term "Euteleostomi" for Sacropterygii and Actinopterygii, while Nelson places Actinopterygii and Sarcopterygii (along with the extinct Acanthodii) in grade Teleostomi. Furthermore, Nelson places the tetrapods (mammals, birds, reptiles, amphibians) in class Sarcopterygii in order to make this a monophyletic group.

Characteristics

Osteichthians are characterized by a relatively stable pattern of cranial bones, rooted teeth, and medial insertion of mandibular muscle in lower jaw. The head and pectoral girdles are covered with large dermal bones. The eyeball is supported by a sclerotic ring of four small bones, but this characteristic is missing or modified in many modern species.

The labyrinth in the inner ear contains large otoliths. The braincase, or neurocranium, is frequently divided into anterior and posterior sections divided by fissure.

Osteichthyans have a lung or swim bladder. They do not have fin spines, but instead support the fin with lepidotrichia (bone fin rays). They also have an operculum, which helps them breathe without having to swim.

The fin-limbs of sarcopterygiians show such a strong similarity to the expected ancestral form of tetrapod limbs that they have been universally considered the direct ancestors of tetrapods (four-legged vertebrates) in the scientific literature.

Replacement Bone

One of the best-known innovations of the osteichthians is endochondral bone or "replacement" bone, i.e. bone ossified internally, by replacement of cartilage, as well as perichondrally, as "spongy bone."

In the more general vertebrates, there are various types of calcified tissue: dentine, enamel (or "enameloids"), and bone, plus variants characterized by their ontogeny, chemistry, form and location. But endochondral bone is unique because it begins life as cartilage.

In more basal vertebrates, cartilaginous structures can become superficially calcified. However, in osteichthians, the circulatory system invades the cartilaginous matrix. This permits the local osteoblasts (bone-forming cells) to continue bone formation within the cartilage and also recruits additional, circulating osteoblasts. Other cells gradually eat away at the surrounding cartilage.

The net result is that the cartilage is replaced from within by a somewhat irregular vascularized network of bone. Structurally, the effect is to create a relatively lightweight, flexible, "spongy" bone interior, surrounded by an outline of dense, lamellar periostial bone. Since this bone now surrounds other bone, rather than cartilage, it is referred to as periostial rather than perichondral. This is the unique endochondral bone from which the osteichthians derived their name, as well as many structural advantages.

However useful endochondral bone may be, it is also much heavier and less flexible than cartilage. Thus, many modern osteichian groups, including the extremely successful teleosts, have evolved away from extensive use of endochondral bone.

Examples:

The ocean sunfish is the most massive bony fish in the world (but not the longest one; that honor goes to the oarfish). Specimens of ocean sunfish have been observed up to 3.33 m (11 ft) in length and weighing up to 2,300 kg (5,070 lb). Other very large bony fish include the Atlantic blue marlin, some specimens of which have been recorded as in excess of 820 kilograms (1,807.4 lb.), the black marlin, and some sturgeon species.

Sarcopterygii

Sarcopterygii or lobe-finned fishes is a type of bony fish that possess lobulated caudal and pectorals fins. For this reason they are called legged–fish.

Taxonomy

The taxonomy of these fish is divided into three subclasses. In the first classification we find the Actinistia which contains the coelacanths: The western Indian Ocean and the Indonesian coelacanth.

The Dipnoos classification, also known as lungfish, is a subclass of freshwater fish, known for possessing primitive characteristics within the bony fish group, including the skill to breathe air, and primitive structures within lobular fin fish, including the presence of lobed fins with a well-developed internal skeleton.

Finally there are the Tetrapodomorpha, (tetrapods) and their extinct relatives. They are a class of vertebrates consisting of vertebrate tetrapods of four limbs and their closest sarcopterigia relatives that are more closely related to live tetrapods than to live lungfish.

Characteristics

Their fins, both caudal and pectoral, consist of fleshy muscular lobes. Each lobe is supported by a central nucleus of individual bones articulated to each other. Most of these bones can be related to those of the extremities of terrestrial animals.

The rounded tip of each fin is hardened by bony rays, which open in a fan. The muscles of each lobe can move the fin rays independently. Within this group it stands out as the amphibians' possible predecessor (or at least the first in chronologically terms), those of the order of the Osteolepiformes, (megalichthyiformes), which became extinct 130 years ago.

Most species of fish with fleshy fins are extinct. The largest known finned-fish was the Rhizodus hibberti (Scottish Carboniferous period) that may have exceeded 7 meters in length.

Actinopterygii

Actinopterygii, is a major taxonomic class (or subclass) of fish, known as the "ray-finned fishes," whose diverse number of species includes about half of all known living vertebrates and 96 percent of all fish species. The actinopterygians include the most familiar fish, such as sturgeons, gars, eels, carp, herrings, anchovies, catfishes, goldfishes, piranhas, oarfish, seahorses, bass, cichlids, pickerel, salmon, and trout.

The ray-finned fishes are so called because they possess lepidotrichia or "fin rays," their fins being webs of skin supported by bony or horny spines ("rays"), as opposed to the fleshy, lobed fins characteristic of the Sarcopterygii, which together with the actinopterygians comprise the superclass Osteichthyes, or bony fish. The actinopterygian fin rays attach directly to the proximal or basal skeletal elements, the radials, which represent the link or connection between these fins and the internal skeleton (e.g., pelvic and pectoral girdles).

As a group, the ray-finned fish play a huge role not only in aquatic ecosystems, both marine and freshwater, where they serve as both prey and predator, but also in diverse areas of human livelihood, from commerce to recreation, aesthetics, recreation, and nutrition. These fish provide essential nutrition for millions of people, are viewed in aquariums and underwater, are sought through sports fishing, and play a fundamental role in food chains, ensuring healthy ecosystems and controlling prey populations, including insects. Beyond this, their extraordinary diversity and geographic range—ubiquitous throughout fresh water and marine environments from the deep sea to the highest mountain streams (with some species even venturing outside of water) and with a spectacular array of colors, body forms, and behaviors—has added greatly to the human enjoyment of nature.

Actinopterygii (the plural form of Actinopterygius) is commonly placed as a class of vertebrates, generally with the parent taxon Osteichthyes (the bony fish) listed as a superclass. In some classification schemes, however, Osteichthyes is listed not as a superclass but as a class, in which case Actinopterygii is listed as a subclass. Alternatively, Nelson, while using Actinopterygii as a class, chooses not to use Osteichthyes as a formal taxon at all because it "is clearly not a monophyletic group."

With such a great multitude of species, the Actinopterygians' characteristics tend to be spread over a considerable range. Many, but not all, of the Actinopterygians, for example, have scales, which may be either the more primitive ganoid form(diamond-shaped, shiny, hard, and multilayered), or the more advanced cycloid or ctenoid forms, which overlap a bit like roof tiles from head to tail to reduce drag. Cycloid scales have smooth edges and ctenoid have rough edges. Other characteristics include nostrils set relatively high up on the head with internal nostrils absent, spiracle (a hole behind the eye through which some cartilaginous fish pump water to the gills) usually absent, pectoral radial bones attached to the scapulo-coracoid skeletal complex (except in Polypteriformes), interopercle and branchiostegal rays (bone-like infrastructure) usually present, and bony gular plate (protecting the throat and lower jaw) usually absent. This group is considered to be monophyletic. Extant species can range in size from Paedocypris, at 7.9 millimeters (0.3 inches), to the massive ocean sunfish, at 2,300 kilograms (5,100 lb), and the long-bodied aarfish, to at least 11 meters (36.1 feet).

Most bony fish belong to the Actinopterygii; there are only eight living species of lobe finned fish (class Sacopterygii), including the lungfish and coelacanths. Nelson and Jonna recognize 42 orders, 431 families, over 4,000 genera, and about 24,000 species of ray-finned fish. This is about half the number of species of known extant vertebrates. About 42 percent of the species of bony fish are known only or almost only from freshwater. However, species of fish are not only being continually discovered, but also some are believed to be becoming extinct faster than they can be discovered.

Actinopterygians are generally classified into two groups—the Chondrostei and the Neopterygii. The Chondrostei include paddlefishes, sturgeons, and bichirs. The Neopterygii include Amiiformes (bowfin), Semionotiformes or Lepisosteiformes (gars), and Teleostei (modern bony fishes). Most fish today fit into the Teleostei, with about 23,000 of the 24,000 actinopterygians being teleosts.

Other classifications of the Actinopterygians exist. For example, in addition to Chondrostei and Neopterygii, Lundberg also lists a taxon of Actinopterygii known as Cladistia, comprised of the bichirs, reedfishes, Polypteriformes, and Polypteridae.

Diversity

The ray-finned fish are extraordinarily diverse in terms of body form, color, habitat, behavior, and so forth. They live in almost all types of habitats with the exception of land that is constantly dry (and some species spend a considerable amount of time outside of water), including the depths of the ocean to 7,000 meters, subterranean caves, desert springs and ephemeral pools, high altitude lakes, and polar seas, and including temperatures from -1.8°C to nearly 40°C (28.8°F to nearly 104°F), salinities from 0 to 90 parts per million, pH levels from four to above ten, and dissolved oxygen levels down to zero. Actinopterygians may swim, walk, fly, or be immobile, they feed on nearly all types of organic matter, they exhibit a huge variety of colors, and they have different types of sensory systems, including vision, hearing, chemoreception, electroreception, lateral line sensation, and so forth. The electric eel and various other fish can produce electric organ discharges (EODs), which may be low voltage for electrolocation and high voltage to stun prey or offer protection.

References

- Classification-of-fish: animalsworlds.com, Retrieved 11 March, 2019
- Agnathan, animal: britannica.com, Retrieved 18 June, 2019

- Hardisty, M. W.; Potter, I. C. (1971). Hardisty, M. W.; Potter, I. C. (eds.). The Biology of Lampreys (1 ed.). Academic Press. ISBN 9780123248015

- Hagfish: newworldencyclopedia.org, Retrieved 11 January, 2019

- Chondrichthian, animal: britannica.com, Retrieved 5 May, 2019

- Osteichthyes: newworldencyclopedia.org, Retrieved 17 July, 2019

- Sarcopterygii, c-fishes: ourmarinespecies.com, Retrieved 8 February, 2019

- Actinopterygii: newworldencyclopedia.org, Retrieved 29 April, 2019

Chapter 3

Fish Anatomy

The anatomy of fish is divided into two broad categories, namely, internal and external anatomy. The external fish anatomy includes fins, skin, gills, eyes and mouth. The components of internal fish anatomy are brain, swim bladder, kidney, stomach and heart. The chapter closely examines these key components of fish anatomy to provide an extensive understanding of the subject.

Fish anatomy is primarily governed by the physical characteristics of water, which is much denser than air, holds a relatively small amount of dissolved oxygen, and absorbs light more than air does. Nearly all fish have a streamlined body plan, which is divided into head, trunk and tail, although the dividing points are not always externally visible.

The head of a fish includes the snout, from the eye to the forwardmost point of the upper jaw, the operculum or gill cover, and the cheek, which extends from eye to preopercle. The lower jaw defines a chin. The head may have several fleshy structures known as barbels, which may be very long and resemble whiskers. Many fish species also have a variety of protrusions or spines on the head. The nostrils or nares of almost all fishes do not connect to the oral cavity, but are pits of varying shape and depth.

The outer body of many fish is covered with scales. Some species of fish are covered instead by scutes. Others have no outer covering on the skin; these are called naked fish. Most fish are covered in a protective layer of slime (mucus).

The lateral line is a sense organ used to detect movement and vibration in the surrounding water. It consists of a line of receptors running along each side of the fish. The caudal peduncle is the narrow part of the fish's body to which the caudal or tail fin is attached. The hypural joint is the joint between the caudal fin and the last of the vertebrae.

Skeletal System

The skeletal system forms the framework of the body and is mainly composed of the bones and tissues. The main functions of the skeletal system are as follows:

- Protection,
- Helps in movement,
- Provides shape and support,
- They are the site for the storage of minerals like calcium.

Fish is an aquatic organism which belongs to the subphylum Pisces. In the taxonomic hierarchy, fishes belong to the kingdom Animalia, phylum Chordata. They are in diverse groups which include jawless fish, armoured fish, cartilaginous fish, lobe-finned fishes and ray-finned fish and so on. Among these, most of them are ectothermic i.e. cold-blooded organisms. According to the behavior and characteristics of nature, they show diversity in their bodily structures. Most of the fishes have strong and firm bony skeletal system.

Skeletal System of Fish

General features of fishes include fins, streamlines and scales and tails. But differences are highlighted under their skins. Hence classification is much easier based on the skeletal system. A variety of fishes is found in aquatic habitat some may be cartilaginous (Chondrichthyes) or bony fishes (Osteichthyes). The skeletal system of fishes is either composed of thin and flexible cartilage or hard calcified bones or both. They are good swimmers and their body structures are designed accordingly.

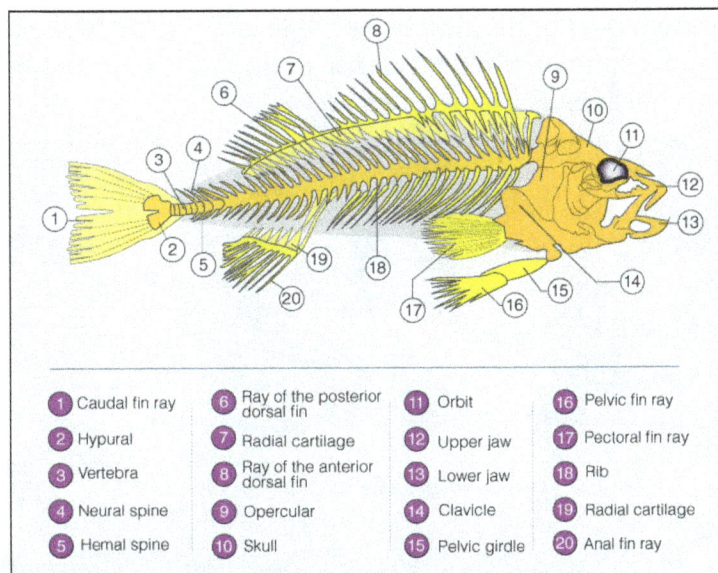

1 Caudal fin ray	6 Ray of the posterior dorsal fin	11 Orbit	16 Pelvic fin ray
2 Hypural	7 Radial cartilage	12 Upper jaw	17 Pectoral fin ray
3 Vertebra	8 Ray of the anterior dorsal fin	13 Lower jaw	18 Rib
4 Neural spine	9 Opercular	14 Clavicle	19 Radial cartilage
5 Hemal spine	10 Skull	15 Pelvic girdle	20 Anal fin ray

Skeletal System of Fish.

Composition of Fish Skeletal System

A skeleton of fish is either made of bone or Cartilage. There are two different skeletal types:

• Exoskeleton- An outer shell of an organism.

• Endoskeleton- Inner shell of an organism.

The main features of the fish skeletal system are it consists of the vertebral column, jaw, ribs, cranium and intramuscular bones. It provides protection and control and also they produce red blood cells in addition to kidneys and spleen. Starting from the head, bony fish consist of solid hard bones called cranium. Cranium protects the brain from mechanical stresses. Osteichthyes have hinged jaws which aid them in feeding. But fishes like hagfish, lampreys are jawless fishes. Swim bladder helps them to take up the dissolved oxygen from water and provides buoyancy. Otoliths are unique characteristics of ear plates of bony fish which helps in steadiness.

Gills are the pair of the respiratory organ of fish and some amphibians. There are three pairs of bones that aid gills. These pair of bones is called gill arches which are made of bony filaments. Fins are a vital part of fish. They help with propulsion, steering, and stability. Paired fins take up the role of steering while caudal fins and dorsal fins help in propulsion and stability respectively.

External Fish Anatomy

Fins

Fins are one of the most distinguishing features of a fish and they have several different forms. Two types of fins are found in most of the fish: median and paired fins. Median fins are single in number which runs down the mid-line of the body. In fishes, median fins are dorsal, caudal and anal fins while paired fins are pectoral and pelvic which are arranged in pairs homologous to human arms and legs. Fins help to swim and maintain the balance of the body. Fins also help to identify the fish species. Different types of median and paired fins are:

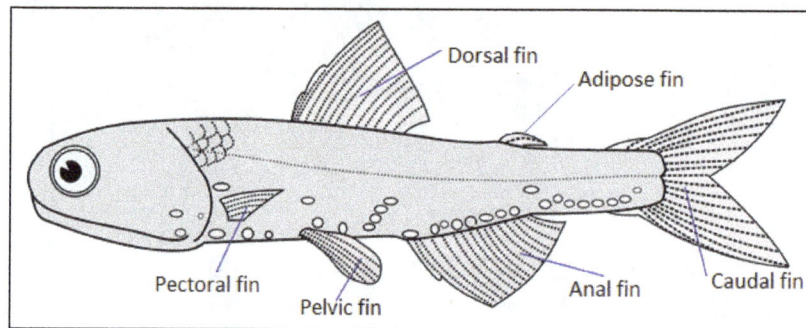

Fish showing different types of fins.

Dorsal Fin

This type of fin is located on the top or back of the fish which help the fish in quick turns or stops. It also helps the fish against rolling. In fish, there are three distinct dorsal fins such as proximal, central or middle, and distal dorsal fins. Some fish have two dorsal fins where the central and distal fins are combined together.

The Types of Dorsal Fins are:

- Single
- Pointed
- Split
- Spine triangular
- Trigger
- Trailing

Pelvic Fin

In fishes, pair of pectoral fins is present which are located ventrally below and behind the pectoral fins. In some fishes, they are situated in front of the pectoral fins (Cod family). This type of fin helps in stability and slowing down the fish. Generally, fish use pelvic fins for moving upwards and downwards in the water.

Anal Fin

The Anal fin is also known as cloacal fin which is located on the ventral side just behind the anus. It supports the dorsal fin and stabilizes the fish during swimming and contrinols the rolling motion.

Pectoral Fin

Pectoral fins are located on both sides usually just behind the operculum. It is homologous to the tetrapod`s forelimbs. It provides supports during swimming. It creates dynamic lifting force and also helps the fish to turn left or right.

Adipose Fin

They are soft fins and located between the dorsal and caudal fins, usually very near to the caudal fin. It is mainly found in Salmonidae, Characins, and catfishes. This type of fin helps to navigate the fish in rough water.

Caudal Fin

The caudal fin is the primary appendage which is used for locomotion in many fishes. The caudal fin is also known as tail fin or a median fin which is usually homocercal or heterocercal. Generally, it is a vertically expanded structure which is located at the caudal end of the body. The base of the caudal fin is known as caudal peduncle with strong swimming muscles. In general, caudal fin acts like a propeller while the caudal peduncle functions as a motor.

The caudal fin has two lobes such as dorsal epichordal and ventral hypochordal lobe which are supported by the modified last three caudal vertebrae. The shape of the caudal fin may vary in different species from rounded to pointed, notched, emarginated, truncated, etc. It is used to identify the fish species. Generally, fish use it for forwarding propulsion and speed.

The caudal fin of the adult fishes may be grouped into three categories:

- Protocercal Caudal Fin: It is the most primitive type of caudal fin where the straight vertebral column divides the caudal fin into two equal lobes such as upper lobe and lower lobe. In this case, the upper lobe is known as epichordal or epicaudal and the lower lobe is called hypochordal or hypocaudal lobe. A series of rods are arranged around the central axis of the caudal region, which support the fin membrane. Undoubtedly, during the developmental period, the caudal fin of all fishes passes through the protocercal stage. This type of fin is found in cyclostomes and the living dipnoans (lungfishes).

- Heterocercal Caudal Fin: The heterocercal tail is sometimes called the shark-tail type of caudal fin.Elasmobranch (cartilaginous fish) and some primitive type of bony fishes contains this type of fin. This fin has two unequal lobes where the upper smaller lobe is known as epichordal lobe and a much larger lower lobe is known as hypochordal lobe. In this case, the hind end of the vertebral column becomes bent upwards and continues almost up to the tip of the fin.

- Homocercal Caudal Fin: Most of the higher teleosts have homocercal caudal fin. It has superficially symmetrical and two equal sized lobes such as upper epichordal and the lower hypochordal lobe. Internally, this tail is asymmetrical and the hinder end part of the vertebral column is greatly shortened and turned upward. In this case, the vertebral column does not touch the posterior limit of the fin.

Different Types of Caudal Fins

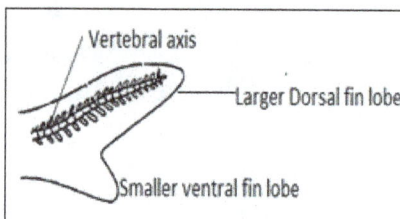

Heterocercal fin.

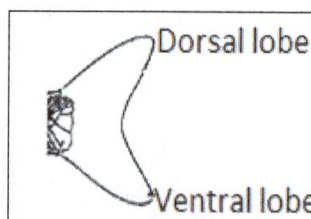

Homocercal fin.

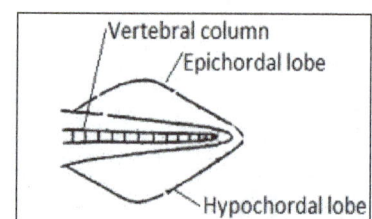

Dephycercal fin.

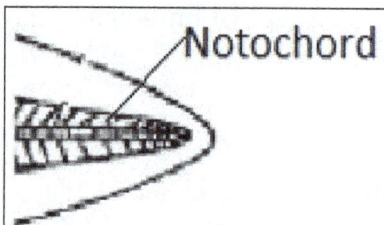

Protocercal fin.

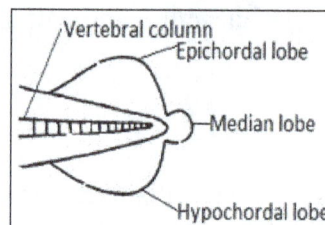

Modified Dephycercal fin.

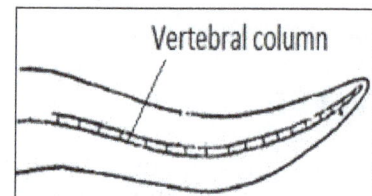

Isocercal fin.

Varieties of Caudal Fins

The internal and external structure of caudal fin varies which depends on the swimming habits of the fish. Generally, these variations involve special modification of the vertebral column. Following seven main types of caudal fins are found in fishes:

- Lunate or Crescentic: It is used for Continuous long distance swimming. E.g. Tuna.

- Forked: It is used for rapid swimming, e.g. Herring, Mackeral.

- Emarginate: e.g. Trout, Carp, Perch.

- Truncate: It aids in turning quickly. E.g. Flounder.

- Rounded: It is used for slow swimming, accelerating, and maneuvering. E.g. Turbot and Lemon-Sole.

- Pointed: e.g. Gobby.

- Double emarginated.

Caudal Fins Shapes

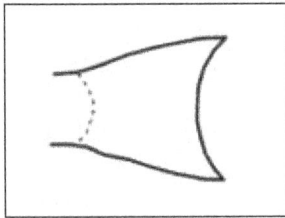
Emerginate

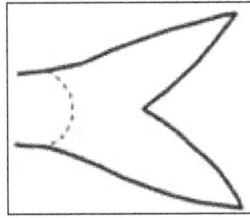
Forked

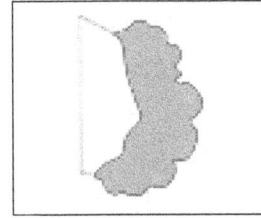
Gephycercal

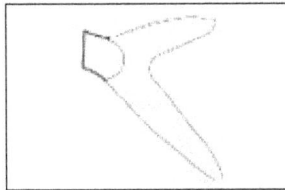
Hypocercal

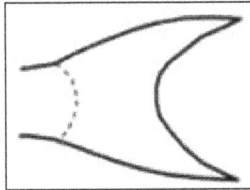
Lunate

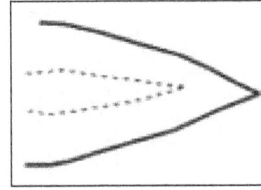
Pointed

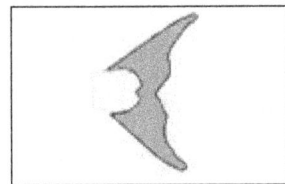
Double emerginate

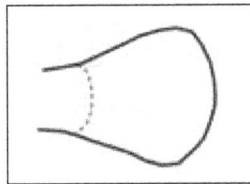
Rounded

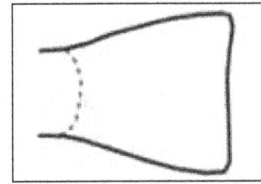
Truncate

Modifications of Caudal Fin

Many fishes have specialized modified types of caudal fins such as:

- Isocercal or leptocercal

- Internally symmetrical Caudal Fin

- Pseudocercal Caudal Fin

- Hypocercal Caudal Fin

- Gephyrocercal Caudal Fin

1. Isocercal or Leptocercal: In some fishes, tapering and symmetrical types of fin is present which is known as isocercal or leptocercal caudal fin. In this case, the spine is long with a straight rod-like structure. Rat tails (Macruidae), Blennies (Blennidae) Eels (Anguilliformes), feather backs (Notopteridae), and Gymnarchids (Gymnarchidae), etc have the isocercal caudal fin.

2. Internally Symmetrical Caudal Fin: This is a reduced type of caudal fin where some fin elements are fused together. They are found in cods (Order: Gadiformes).

3. Pseudocaudal Caudal Fin: In the modern lungfishes (Dipnoi), the pseudocaudal caudal fins are found. In this case, fins are developed from the backward growth of dorsal and ventral elements.

4. Hypocercal Caudal Fin: This type of caudal fin bears much larger dorsal lobe than the ventral lobe which is greatly reduced. They are found in certain early Agnathans. It is also known as inverted heterocercal caudal fin. In this case, the vertebral axis turns downwards sharply where the lobe develops from its upper surface.

5. Gephyrocercal Caudal Fin: It is a very specialized type of caudal fin which is also known as bridge caudal fin. Generally, they look like the isocercal fin but the fins are reduced to vestiges. In this case, the caudal lobe is truncated where hypurals of the spinal column are lacking. These types of fins are found in the pearlfishes (Carpus), Flerasfer and Orthagoriscus.

Functions of Fins

Fish use their fins for various purposes. Some important functions of fins are described below:

- Generally, the pectoral fins help a fish for turning.

- Some bony fishes use their pectoral fins to help them rest on the bottom or on reef areas (e.g. Cirrhitichthys).

- Mudskippers (Periophthalmidae family) use pectoral fins for supporting themselves on land.

- Flying fish (Exocoetidae family) use their long pectoral fins for gliding over the water.

- Pectoral fins of some bottom-residence fishes such as threadfins (Polynemidae) bear touch receptors and taste buds which help to trace food.

- Pelvic fins help the fish stability in the water.

- Pelvic fins of some fishes such as clingfishes (family Gobiesocidae) use as sucking appendage, which helps a fish hold on to stationary objects on the ocean bottom.

- Most of the bony fishes use their dorsal fin for sudden direction changes.

- Dorsal fins act as a 'keel' for keeping the fish stable in the water.

- Some angelfishes (Lophiiformes) use their dorsal fin as a lure which helps to attract the prey.

- The modified dorsal fin of some fishes (Echeneidae) use as a sucking disc.

- African knife fish (Gymnarchus niloticus) use its dorsal fin to move forward or backward by creating undulating.

- Most of the bony fish use their caudal fins for propulsion.

- Lunate caudal fins are characteristic features of fast swimmers such as tunas. They use it for maintaining rapid speed for long duration.

- Anal fins make stability and anal fins of some bony fishes help in reproduction.

- Sea Robin fish use their pelvic fin for walking along the substrate.

- Some fishes such as Freshwater butterflyfish (Pantodon buchholzi) use their pelvic fins for gliding.

- Sea Robins use their pectoral fin for gliding around in the currents.

- Lionfish and other scorpionfish have dorsal fins with hollow venomous spines which are used for self-defense.

Skin

The integument or skin is an outermost covering or wrapping of the body, hence it is the most exposed part of the body to the environment. For this reason, it plays an important role of first line of defence in a number of ways. In fishes, the skin is well-adapted for protection from injuries and diseases. It also serves for respiration, excretion and osmoregulation.

In some fishes, special colouring devices and phosphorescent organs are present in the skin, which either conceal the organism or make it present or used for sexual recognition. In addition, some species have special structures like electric organs, mucous glands and poison glands.

Structure of Skin of Fishes

The skin of fish is made up of two distinct layers, viz. an outermost layer, the epidermis and an inner layer dermis or corium. The epidermis originates from ectoderm and the dermis derives from mesoderm layer.

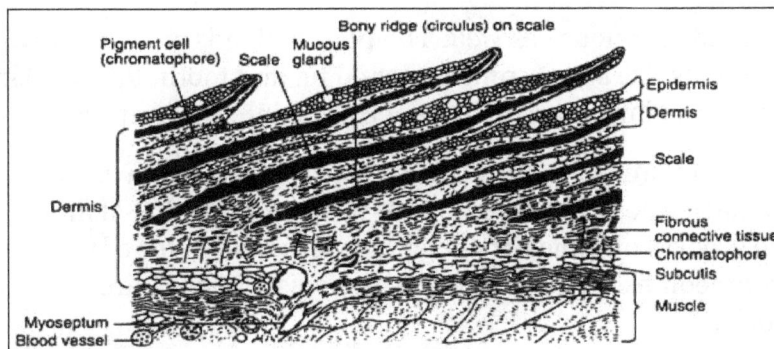

Section of fish skin.

The Epidermis Layer of Skin in Fishes

It is composed of many layers of flattened and moist epithelial cells. The innermost layer is called stratum germinativum. This layer is made up of active columnar cells that continuously divide by mitotic division. The newly formed cells occupy the lowermost stratum and the older cells move outside and are worn off time to time and maintain growth. These migrating epithelial cells fill the superficial wounds.

Epidermal Glands

The epithelium of epidermis is modified into a variety of glands, which are:

- Mucous gland:

    The epidermis is provided with a number of mucous glands, which open at the surface of the skin by minute pores. These glands are flask-shaped or tubular extending to the dermis. The mucus glands secrete slippery mucus, which contain a lipoprotein, known as mucin.

The slimy mucus reduces drag on fish while swimming in the water. Continuous secretion and sloughing of mucus wash away micro-organism and irritants, which may cause disease if accumulated. In some species (Protopterus and Lepidosiren), the mucus forms a cocoon-like structure around the body to avoid dry condition of weather, especially during aestivation. The mucus gives a characteristic fish odour.

Among some fishes, mucous is used for chemical communication. Many teleosts feed their young ones on the mucus, secreted in large quantity on the surface of the body. Some species like Macropodus and Gasterosteus use their sticky mucus for preparation of nest for laying eggs.

The mucus also helps in regulating to some extent, the osmotic exchange of water and ions between the body-fluids and the water. The number and size of mucus gland cells vary with species. Generally, fishes with no scales have large numbers of mucus cells.

- Poison glands:

Venom or poison glands have evolved in different families of fishes. Glandular cells of epidermis are modified into poison glands. These glands secrete poisonous substance to protect themselves from the enemy for defence.

They are also used for offence as well. The poison glands are generally present at the base of certain structures like sting, spine of dorsal fin and tooth. Poison glands open at the tip of these structures to inject poison by penetration into the prey.

The most common example is the stingray, which is provided with venomous caudal sting. Similarly, Chimaeras possess venom glands in spine of the dorsal fin. The poison glands are present in the grooves of spines of dorsal, pelvic and anal fins of the Scorpion fish (Scorpionidae). In Sturgeon fish (Acanthuridae), the poison glands are found at each side of the caudal peduncle.

- Photophores:

In many marine species of fish, special multicellular glands are developed from stratum germinativum of epidermis. These glands are deeply seated into dermis and produce light. These light producing luminous organs are mostly found in deep-sea elasmobranches and in some teleosts inhabiting total darkness in sea.

Each gland has an apex consisting of mucus cells that helps to magnify light, produced from the basal glandular part of the gland.

The Dermis Layer of Skin in Fishes

The dermis lies beneath the epidermis. This layer contains blood vessels, nerves, connective tissues and sense organs. The upper layer of dermis is made of loose connective tissues and is known as stratum spongiosum, while the lower part is occupied by thick and dense connective tissues, called the stratum compactum.

This layer generally has proteinaceous collagen fibres and mesenchymal cells. The dermis is well supplied by blood vessels; hence it also provides nourishment to the epidermis.

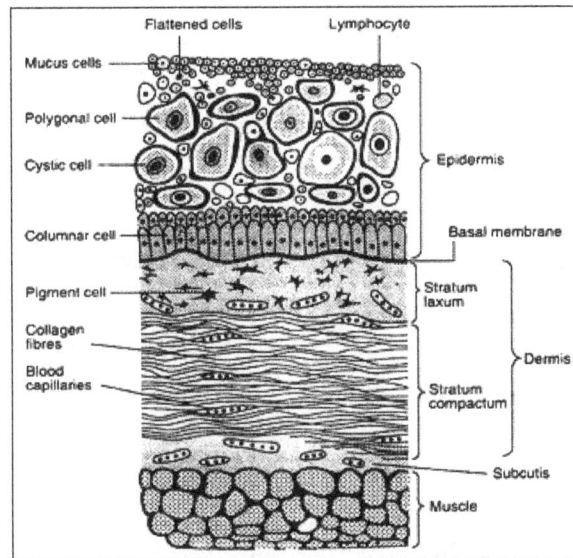

V.S fish skin (generalized).

Scales in Fishes

Scales are derivatives of mesenchymal cells of dermis. Some fishes are "naked" devoid of scales, e.g., freshwater catfish. Certain species exhibit an intermediate condition that are generally naked but possess scales on restricted areas. Such condition is found in paddlefish (Polydon), in which scales are present in region of throat, pectoral and base of tail.

In some fishes, scales are modified into teeth, bony armour plates (Sea horse) and spiny stings (sting ray). In fresh water eel (Anguilla), scales are very small and so deeply embedded that the fish appears to be naked.

Mostly scales are arranged in imbricate manner and overlap with free margin directed towards the tail that minimizes friction with water. In freshwater eel (Anguilla), the arrangement is mosaic; the scales unite their neighbouring ones at their margins.

Type of Scales in Fishes

There are few type of scales based on their structure and shape. The different type of scales are often characteristics of the species.

On the basis of shape, scales are of four types:

1. Plate like or placoid scales commonly found in Elasmobranches.

2. Cycloid scales found in Burbot and soft-rayed fishes.

3. Rhombic or diamond shaped scales, common among gars and sturgeons.

4. Ctenoid scales, characteristics of spiny-rayed bony fishes (Acanthopterygii).

Scales may also be classified as placoid or non-placoid. Three basic types of non-placoid scales are — cosmoid, ganoid and bony-ridge.

Placoid Scales

Placoid scales are characteristically found among the sharks and other Elasmobranches. They are small denticles that remain embedded in the skin. Each scale has two parts, an upper part, known as ectodermal cap or spine. This part is made of enamel, like substance, known as vitreodentine similar to human tooth.

Another layer of dentine that encloses a pulp cavity follows the vitreodentine. The lower part of placoid scale is a disc-like basal plate, which is embedded in dermis with cap or spine projecting out through epidermis.

The basal plate has a small aperture through which blood vessels and nerves enter into pulp cavity. The placoid scales are modified in jaw teeth in sharks; in spines in dorsal fins; in Squalus (spiny dogfish); in sting in the stingrays and in saw teeth in the Pristis.

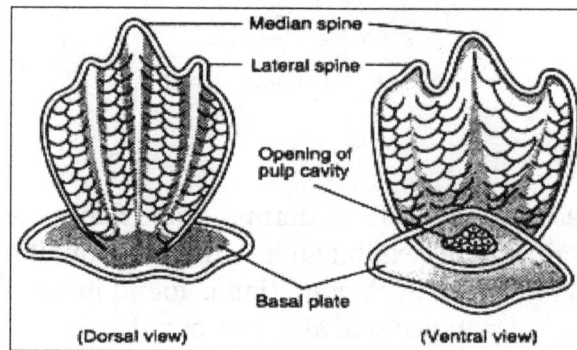

Placoid.

Development of Placoid Scale

The placoid scale first appears as small aggregations of dermal cells just below the stratum germinativum. These dermal cells grow upwards into an arched structure or papilla, which gradually pushes the stratum germinativum. The cells of stratum germinativum of this region become glandular and act as enamel organ.

Later on, this projecting structure differentiates into a spine and a basal plate. The outer cells of the papilla, known as odontoblasts, secrete dentine around the papilla, while the central cells do not calcify and constitute the pulp. The vitreodentine to form a cap over the spine, gradually envelopes the spine of the scale.

The mesenchymal cells of the dermis secrete the basal plate. These cells secrete hard cement-like substance to cover the basal plate. Finally, the spine erupts from the cells of epidermis and projects out, while the basal plate lies embedded in the dermis.

Cosmoid Scale

The cosmoid scales are found in living (Latimaria) and extinct lobefins. In dipnoi, the cosmoid scales are highly modified and appear like cycloid scale. The cosmoid scale is a plate-like structure and consists of three layers! An outermost layer is thin, hard and enamel like, called vitreodentine. The innermost layer is composed of vascularized perforated bony substance, called isopedine.

The middle layer is made up of hard non-cellular and a characteristic material, called cosmine and is provided with many branching tubules and chambers. These types of scales grow at the edges from beneath by addition of new isopedine material.

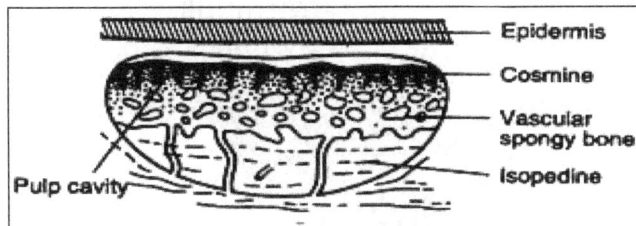

Cosmoid scale (extint crossopterygians).

Ganoid Scale

The ganoid scales are thick and rhomboid. They consist of an outer layer of hard inorganic substance called ganoine, which is different from vitreodentine of placoid scales. The ganoid layer is followed by a cosmine-like layer provided with many branching tubules.

A bony layer of isopedine occupies the innermost layer. These scales not only grow at the edges but also grow at the surface. The growth takes place by the addition of new layers of isopedine.

The ganoid scale is best found in the Polypterus and Lepidosteius. In these fishes ganoid scales are rhombic plate-like, fitting edge to edge and invest the entire body. In Acipencer, the ganoid scales are modified into large bony scutes, arranged into five rows.

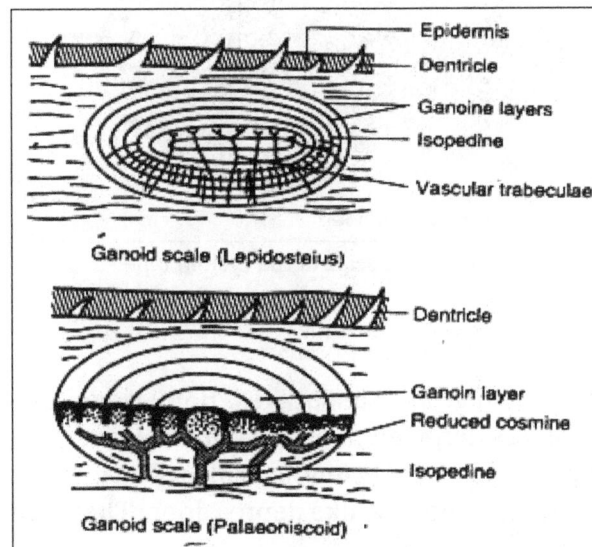

Ganoid scale (lepidosteius).

Ctenoid Scale

They have characteristic teeth at its posterior part. Ctenoid scales are found in spiny-rayed teleost. They are arranged obliquely in such a manner that the posterior end of one scale overlaps the anterior edge of the scale present behind. The chromatophores are present at the posterior part of these scales.

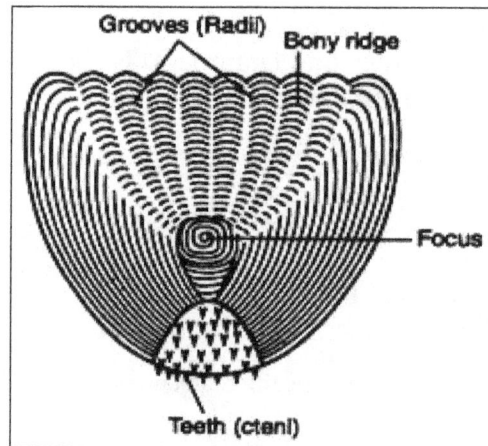

Ctenoid scale.

Cycloid Scale

The cycloid scales are devoid of teeth or spines, hence seem cyclic. They are found in soft-rayed teleost and modern lobe-finned fishes. But some spiny-rayed fishes, i.e., Lepidosteius show presence of cycloid scales. In Micropterus, both cycloid and ctenoid scales are found.

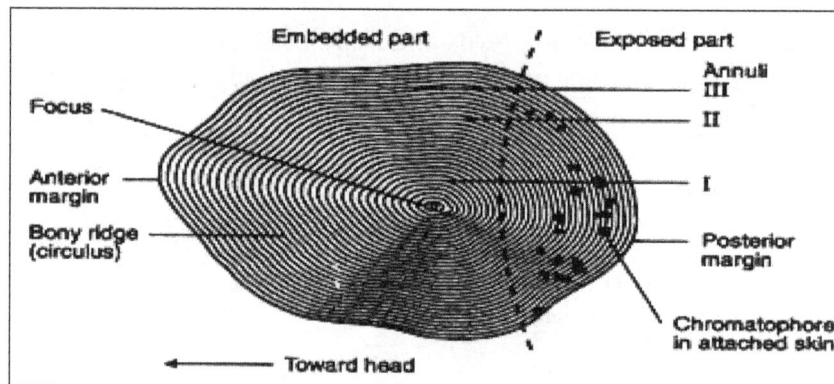

Bony-ridge scale of the cycloid type.

Bony Ridge Scale

Bony ridges characterize the bony fishes, Osteichthyes. Bony ridge scales are thin and semitransparent because they do not possess dense enamel and dentinal layers that are found in other types of scales. They are of two types; cycloid and ctenoid scales. The outer surface of these scales possesses bony ridges that alternate with groove-like depressions. The ridges are arranged in the form of concentric rings.

The inner part of the scale is composed of fibrous connective tissue. The central zone of scale is differentiated properly and is known as focus of the scale. During the development, focus appears first and lies in the central position.

When growth of scales takes place in anterior or posterior parts, it causes shifting of the focus anteriorly or posteriorly, respectively. Later the groves radiate from the focus towards the margin of the scales.

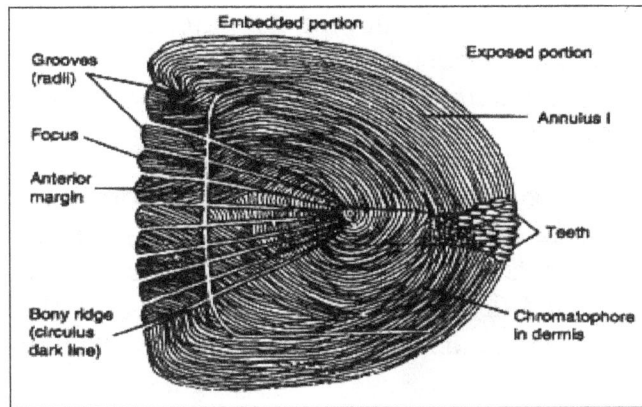

Bony-ridge scale of the ctenoid type.

Development of Bony Ridges

The bony ridges first make their appearance in the dermis as small accumulation of cells at the caudal peduncle and then gradually spread from there. Soon a focus is formed in the centre of the accumulation of cells.

Later on, ridges or circulus are formed at the surface of growing edge of scale. The deepest part of the scale, the basal plate is made up of successive layers of parallel fibres. Some calcification of this fibrillar plate occurs to strengthen the scale.

Significance of Scales in Taxonomy

Scales play significant role in the classification, hence very useful for the ichthyologists. They are not found in the lampreys and hagfishes; sharks are characterized by presence of placoid scales; primitive bony fishes possess ganoid scales; the higher bony fishes have ctenoid or cycloid scales.

The scales' count is highly important in taxonomy. The number of scales present in lateral line, along and around the body, is specific in every species. The age of fish could be determined by measuring space in annual rings of the scales.

In some species like Atlantic salmon, the scales exhibit the presence of spawning marks on them. These marks indicate how many times the fish has spawned and the time of first spawning also.

Gills

Fish breathe through gills instead of lungs. Just like all other animals, fish need oxygen to survive. Because they live in water, they have evolved gills which enable them to remove dissolved oxygen from water. Most fish have four gills on both sides of their head. Sharks and other more primitive fish may have five or more gill slits.

Each gill is supported by a gill arch – a bony structure that is oriented vertically on the side of a fish, just behind its head. The gill arch provides the support to hold a number of comb-like structures called gill filaments. Gill filaments extend out horizontally from the gill arches. Each gill filament produces many branches called primary lamellae and the primary lamellae branch out into tiny secondary lamellae.

The secondary lamellae run parallel to the flow of the water and absorb oxygen from the water into the fish's body. The number of secondary lamellae can vary immensely between fish species but they are always extremely abundant. For example, a large, active tuna can have more than 5 million secondary lamellae per cm².

As water passes over or is pumped over the gills, oxygen is absorbed by through the walls of the secondary lamellae and CO_2 is released. The secondary lamellae contain blood with low levels of oxygen. As water flows over the lamellae oxygen is asborbed into the blood and then the blood pumped around the body by the fish's heart. The large surface area of the secondary lamellae is also helpful for exchanging body heat, ions and water between the fish's body and the surrounding water.

Having so many tiny secondary lamellae creates an enormous surface area for oxygen to be absorbed through. This is helped further by the fact that secondary lamellae have thin walls so gas can be absorbed into the blood stream easier. Dissolved oxygen is found in much lower concentrations in water than it is in air so gills need to be far more efficient with their absorption than lungs do.

Structure of Gills

Gill Slits

There are six or seven pairs of gills in cartilaginous fishes while four pairs in bony fishes due to the loss of spiracle.

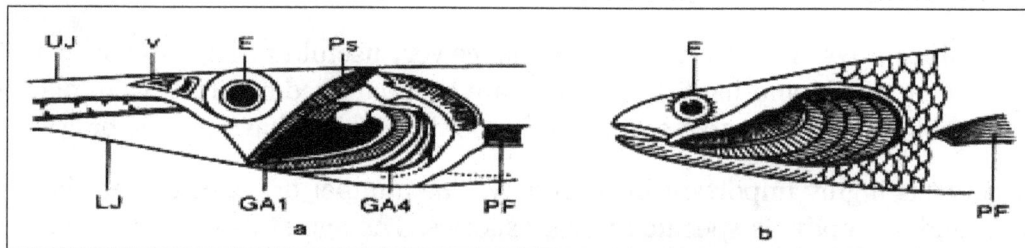

Position of pseudobranch and gill arch in fishes (a) Xenentodon.
(b) Structure of gill of teleost. E, eye; G.A-GA4 gills; L.J. lower jaw.PF, pectoral fin; PS, pseudobranch; UJ, upper jaw; V. Value.

Gill slits of bony fishes are covered by operculum while operculum is absent in cartilaginous fishes. In sharks gill slits are laterally situated while in rays they are ventrally placed. A pair of spiracle is present in Elasmobranchii anterior to first gill which corresponds to a vestigeal primitive first gill slit.

Although spiracle is absent in bony fishes, in Actinopterygii it is replaced by a pseudo-branch which is free in some fishes but skin covered in others.

Pseudo Branch

In carp and rainbow trout the pseudo branch is embedded in submucosal connective tissue of pharyngeal wall and shows a glandular appearance due to complete conglutination of branchial filaments.

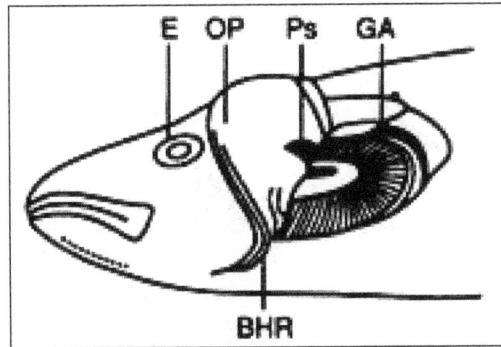

Location of pseudobranch in European walleye (stizostedion lucioperca).
BHR, branchiostegal ray; E, eye; GA, gill arch; OP operculum; Ps, pseudobranch.

In some species, a pseudo branch with hemibranchs structure is located inside the operculum. However, in eel the pseudo branch is not present, it is also absent in cat fishes (Siluroidae) and feather back (Notopteridae).

In glandular pseudo-branch, abundant distribution of blood capillaries is found in the parenchyma enclosed by connective tissue. It contains acidophilic cells in mitochondria and endoplasmic reticulum and is rich in enzyme carbonic anhydrase. The pseudo-branch regulates the flow of the arterial blood to the opthalmic artery to increase the amount of blood carbon dioxide.

Parry and Holliday found that in rainbow trout extirpation of pseudo-branch induced melanophore expansion and body colour change, suggesting the secretion of a melanophore-aggregating hormone from tissue. It also helps in metabolic gas exchange of retina and filling of gas bladder. Because of its direct vascular connection with the choroid gland on the eyeball, the pseudo-branch has been implicated in the regulation of intracellular pressure.

The structure of gills has been studied extensively in fishes by light transmission and scanning electron microscopy. The gill comprises of gill rakers, gill arch, gill filaments (Primary gill lamellae and lamellae).

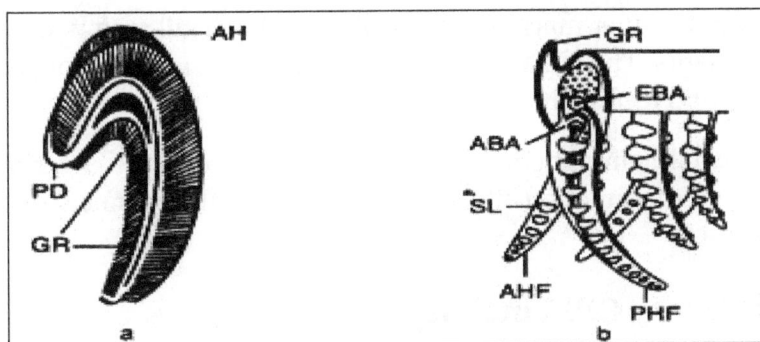

Structure cf gill. ABA, afferent branchial artery, AH, hemibranch;
AHF, aboral hemibranch; EBA, efferent branchial artery, CR, gill raker, PD,
postero dorsal part; PHF, posterior hemibranch filament; SL, secondary lamella.

A complete gill is known as holobranch. It consists of bony or cartilaginous arches. The anterior and posterior part of each gill arch possesses plate-like gill filaments. Each holobranch consists of an anterior (oral) and a posterior (aboral) hemi-branch.

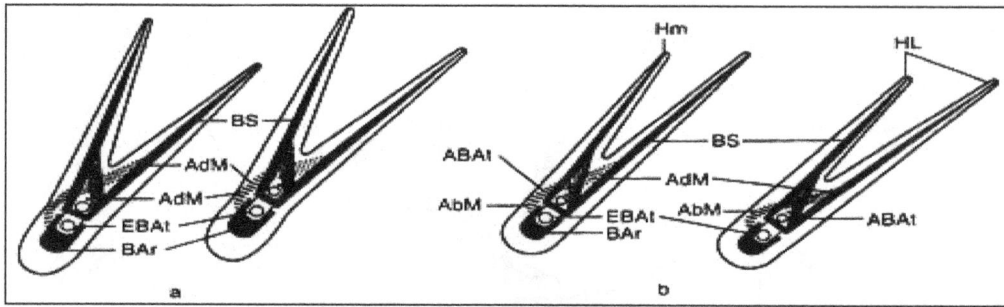

In figure, section of adjacent holobranch and their muscles in bony fish (a) during abduction (b) during adduction. ABAt, afferent branchial artery, AdM, abductor muscles, AdM; adductor muscles; BAr, bony gill arch; BS, bony support; EBAt, efferent branchial artery, HL holobranch; Hm, hernibranch.

The architectural plan of teleostean gills shows heterogeneity in their functional unit which is due to varied osmoregulatory, feeding and respiratory behaviour and to the physicochemical status of their environment.

In teleost fishes, five pairs of branchial arches are present of which first four bear gill lamellae but the fifth is devoid of gill lamellae and transformed into the pharyngeal bone for mastication of food. It does not play any role in respiration.

The gill arch is an important unit and bears primary (gill filament) and secondary lamellae. The branchial arch typically consists of paired pharyngobranchials, epibranchials, ceratobranchials, hypo-branchials and a median unpaired basibranchial.

The epibranchial and the ceratobranchial elements of each branchial arch bear two rows of gill filaments of the two hemibranchs of the holobranch, which are the seat of gaseous exchange. It encloses afferent and efferent branchial arteries and veins.

Gill Raker

It occurs in two rows on the inner margin of each gill arch. Each gill arch is short stumpy structure supported by bony elements. The gill arch projects across the pharyngeal opening. They are modified in relation to food and feeding habits.

The mucous cells of the epithelium help to remove sediments from the covering epithelium in order to enable the taste buds to function effectively and to sense the chemical nature of food passing through the gill sieve.

Gill Filaments (Primary Gill Lmellae)

Each hemi-branch consists of both primary and secondary lamellae.

The primary gill filaments remain separated from the branchial septum at their distal end making two hemi-branches in opposition which direct the water flow between the gill filaments. Amongst dual breathers the heterogeneity in the gill system is more pronounced particularly in the swamp eel, Monopterus, Amphipnous cuchia and climbing perch, Anabas testudineus.

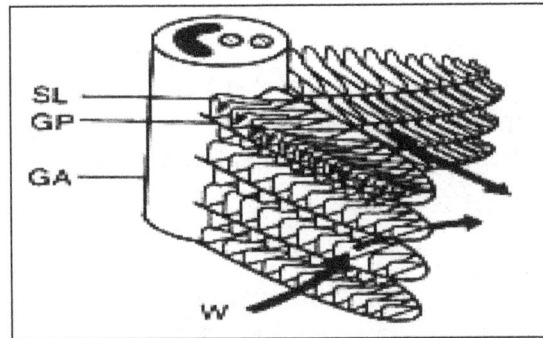

Diagram of gill structure of fish. B, blood flow; G.A, gill arch;
GP, gill filament; SL, secondary lamella; W, water.

In Monopterus, gill filaments are stumpy and are present only in second pair of gill and lack gill lamellae, the remaining three pairs are without functional lamellae. It is the modification for another way of exchange of gases.

The gill filaments are blade-like structures supported by gill rays. The heads of the gill rays of both the hemibranchs are connected by ligaments. They are provided with two types of adductor muscle units in teleosts. The gill filaments are also lined by epithelium referred to as primary epithelium. The epithelium has glandular and non-glandular part.

Lamellae (Secondary Lamella)

The each gill filament is made up of secondary gill lamellae which are actual seat of exchange of gases. They are generally semicircular and lined up along both sides of the gill filaments. The lamellae frequency is directly proportional to the dimension and resistance of the gill sieve.

The secondary lamellae are having two sheets of epithelium which are separated by space and through these spaces blood circulates. The epithelial sheets are separated by a series of pillar cells. Each cell consists of central body and is provided with extensions at each end.

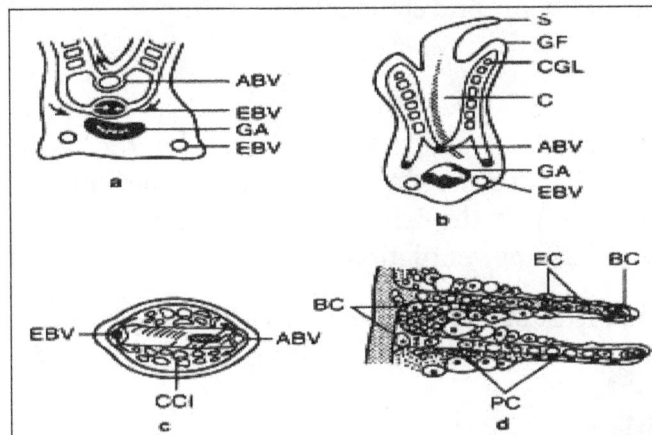

In figure, horizontal section of gills of (a) shark, (b) bony fish. (c) cross-section,(d) epitheliumof gill. ABV, afferent blood vessel; AC, acidophil cells; BC, blood cells; C, cartilaginousgill support; CLL capillaries and lacunae in lamella; CGL, capillary in gill lamellae; EC, epithrlial cells; EBV, efferent branchial vessel; GA, gill arch; GF, gill filament; PC, pilaster cells; S, septum.

Branchial Glands

These are specialized cells of the epithelium. They are glandular in nature and perform different functions in normal and experimental conditions. The most common specialized branchial glands are the mucous glands and acidophilic granular cells (chloride cells).

Mucous Glands

These gland cells are unicellular. They may be oval or pear shaped with a neck through which they open outside the epithelium. The nucleus lies at the bottom of the cells. They are typical goblet cells. They are present throughout the epithelium, i.e., gill arch, gill filament and secondary lamellae.

They secrete mucus which is glycoprotein, both acidic and neutral. Acidic and neutral glycoproteins are secreted by same cells suggesting that they undergo transformation from each other.

The function of mucus cells are as follows:

1. Protective

2. Friction reducing

3. Anti-pathogen

4. Help in ionic exchange

5. Help in gas and water exchange

Chloride Cells

These cells contain granules, which take acidic stain, hence called acidophlic. They are provided with a large number of mitochondria. These cells are also known as ioncytes because they help in the ionic regulation in the euryhaline and stenohaline fishes.

Taste Buds

They are flask shaped multicellular structure and distributed in the epithelium of the gill arch region. There are two distinct types of cells, sensory and supporting, and can be distinguished under phase contrast microscope. They help the fish to sense the nature of food particles contained in the water flowing through the gill during ventilation.

They are distinguished into MGO, MREC and taste buds. The distribution and density of MGO and taste buds on the gill arch epithelium of fishes inhabiting different water bodies differ considerably.

Respiratory Mechanism

The continuous undulational flow of water over gill surface is accomplished by respiratory pump. It is now unanimously accepted that the respiratory pumps of a teleost consist of buccal cavity and two opercular cavities, caused by movements of bone of arches and operculi, resulting in the pumping action of the system.

In the beginning, the water enters into the mouth by expansion of buccal cavity. The water is then accelerated over the gill by the simultaneous contraction of buccal cavity and cavity contracts, expelling the water out through the opercular opening, the cycle begins again.

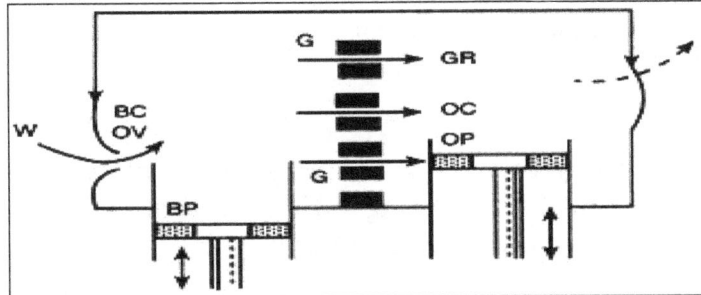

Diagram to show mechanism of gill ventilation by two pumps just after inspiration. BC, buccal cavity, BP, buccal pump; G, gills; GR; gill resistance; OC, opercular cavity; OP, Opening OV, oral valve; W, water.

The respiratory cycle is a complex mechanism and involves a large number of muscles, bones, ligaments and articulations.

Ram Ventilation

It is carried by strong abduction of the two hemibranchs of a holobranch towards each other. Interruption of this cycle causes reversal of flow or cough which fish uses to clear foreign matter or excess mucus from the gills. Frequency of cough in Salvelinus fontinalis can be a sub-lethal indicator of excessive copper concentration in freshwater.

There is active and passive ventilation during the swimming movement. The transition to ram gill ventilation in fish is a graded process as swimming picks up from rest. The first indication that a critical swimming speed has been reached is signaled by the drop-out of a single cycle.

The drop-out continues until only occasionally ventilatory movements and 'cough' are noticed. Return of active movements with gradual reductions in swimming velocity to below critical shows nearly the same sequence but in reverse order.

The additional respiratory mechanism of sharks and rays can be divided into three phases, which are as follows:

1. When coracohyoid and coracobranchial muscles contracts to enlarge the angle enclosed by the gill arches and to increase the oropharyngeal cavity for entrance of water through mouth or spiracle during which the gill slits are kept closed.

2. When the contraction between the upper and lower parts of each gill occurs with the relaxation of abductors of lower jaw and gill arches, which causes mouth to serve as pressure pump. During this phase the forward flow of water through mouth is prevented by the oral valve and the water is directed backward towards the internal gill clefts.

   The inter-septal spaces are enlarged by the contraction of the inter-opercular adductors to reduce the hydrostatic pressure at the inner surface of the gill and water is drawn into the gill cavities, which remain closed at the outside.

3. The third phase includes relaxation of inter-opercular muscle and contraction of their sets of muscle which cause the internal gills to narrow, and the water is forced through the gill lamellae. This follows the opening of gill clefts, and the water is forced to the outside. The mackerel sharks (Limmidae) take the sufficient respiratory water during swimming and do not show pronounced breathing movements.

Mechanism of Entry of Water

In all bony fishes the pressure and flow of water in the oral cavity is regulated by the muscles, which move the bases of holobranchs. "Coughing" is the process of a voilent sweeping of water over the gill lamellae to make them free from accumulated detritus.

It is carried by strong abductions of the two hemibranchs of a holobranch towards each other. The diffusion of oxygen from the respiratory water is added not only by the gill filaments but also by the direction of blood and water flow. These works a counter-current system in which the oxygenated water flows from the oral to aboral side of the gills, but the blood in lamellae flows in opposite direction, i.e., from aboral lamellar afferent to oral lamellar efferent blood vessels. During this counter-flow of oxygenated water and blood, the oxygen enters the gill and carbon dioxide leaves it. In the tench (Tinea) this counter-current system is so modified that the 51% of oxygen from incoming respiratory water diffuses in and 9% oxygen from outgoing water currents.

In the osteichthyes fishes, in which gills are covered by operculum, the water is propelled over the gills by suction pressure. In the beginning of inspiration, when the operculum is closed forcefully, and mouth is opened while several muscles contract including sternohyoid and elevator of palatine. The branchiostegal rays are spread and lowered simultaneously and the mouth cavity is enlarged creating negative water pressure in it. Thus water is drawn into the mouth and after a short interval the space between the gills and the operculum is enlarged as the gill covers are abducted anteriorly although the opercular skin flaps are still closed posteriorly by outside water pressure.

A negative pressure develops in the gill cavity resulting flow of water over gills. This is followed by reduction of buccal and opercular cavities. At this time the oral valves presents the backflow of water out of the mouth and mouth cavity begins to function as pressure pump instead of a suction pump. Then each operculum is immediately brought towards the body, the gill flaps open and water is expelled, being prevented from flowing backwards by high pressure in the buccal cavity as compared to the epibranchial cavity.

Variations in Respiration Pattern

There may be variations in the basic patterns of respiration due to different life habits.

1. Fast Swimming:

    In fast swimming species like mackerels (Scombridae), trouts and salmons (Salmoninae) the mouth and gill flaps remain open to bathe their gills with water currents produced during swimming. Generally, the fast swimmers have smaller gill cavities in comparison to the sedentary fishes.

2. Bottom Dwellers:

The bottom dwelling fishes such as flounders (Plueuronectidae) and gossefishes (Lophiidae), possess enlarged and dilatable opercular cavities, their mouth does not open widely during inspiration causing slow and deep breathing movements. The other bottom dweller species like morays (Muraenidae) keeps its mouth open during the respiration.

Some fishes such as the bonefish (Albula vulpes) and goalt fishes (Mullidae), when find their prey covered by sand grains, they release jet of water from the mouth with the help of strong adduction of gill covers which uncover their food or prey.

The trunkfishes (Ostraciidae) and the puffers (Tetradontidae) possess rather a compact skeleton which reduces respiratory function of the gill cover. However, these fishes have a compensatory high rate of exchange of respiratory water carried out by fast breathing (up to 180 breathing movement/min).

3. Hill Stream:

In some hill stream catfishes who attach their body temporarily to the substratum, the water intake is accomplished by developing grooves protected by barbles with slight reduction in the force of suction. Ventilation is carried out only by the opercular movement.

In the andes (Arges), while attaching to the substratum with their suctorial mouth, catfishes can withdraw and eliminate respiratory water via an inhalent slit in the horizontally divided gill cover. The larvae of South American lungfish (Lepidosiren) and the reed fishes possess "external gills" coming out through external gill slits or opercular region.

Fish Blood as Gas Carrier

Like other vertebrates, the fishes also possess R.B.C. containing respiratory pigment haemoglobin which has efficient oxygen carrying capacity. The haemoglobin has 15 to 25 times more capacity than water to bind with the oxygen. The 99% of total oxygen is taken by haemoglobin while only 1% by the plasma. Haemoglobin is a conjugated protein, the chromo-protein.

It consists of a large protein molecule, the globin consisting of four polypeptide chains (two alpha and two beta chains), to each of which is attached a prosthetic haem group. Haem is based on a structure known as a porphyrin ring which includes four pyrrole groups around a central ferrous iron (Fe^{++}).

The iron is joined by four of its coordination bond to N atom of the porphyrin and two bonds to imidazole N contained in histidine residues within the protein globin.

The haemoglobin contents may vary according to habit and habitate for instance, the pelagic species have higher Hb than bottom dwellers. When haemoglobin is oxygenated, i.e., 'loading' and when transported and released, i.e., 'unloading' this process is oxygenation but not the 'oxidation'.

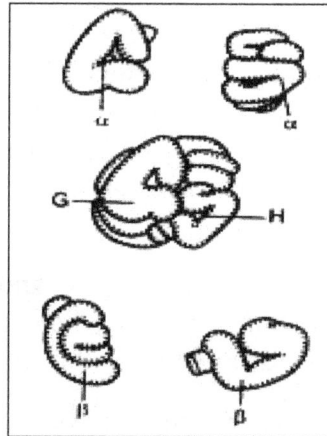

Structure of haemoglobin. α, alpha chain;
β beta chain; G, globin; H, heam.

The loading is also denoted by TI or T, sat, i.e., loading tension of blood which shows that partial pressure of oxygen at which Hb of a particular species is 75% saturated with oxygen.

The Dissociation Curve of Oxyhaemoglobin

The relationship between the saturation of haemoglobin and oxygen tension is studied by examination of the dissociation curve of hemoglobin in which the present saturation is plotted against the oxygen tension. The shape of dissociation curve varies with the tension of CO_2.

The curve may be hyperbolic as found in the eel (Anguilla). This hyperbolic curve with high O_2 affinity makes the fish enable to live in water with low oxygen concentrations. The haemoglobin of eel becomes saturated at much lower tension than that of mammals.

The amount of oxygen given to the tissues can be determined by the difference between the Tsat and T1/2 sat. In contrast to the hyperbolic curve, the sigmoid curve (human) shows efficiency of the blood to release more oxygen to the tissues. This can be exemplified by the curve obtained in more active fish, the mackerel.

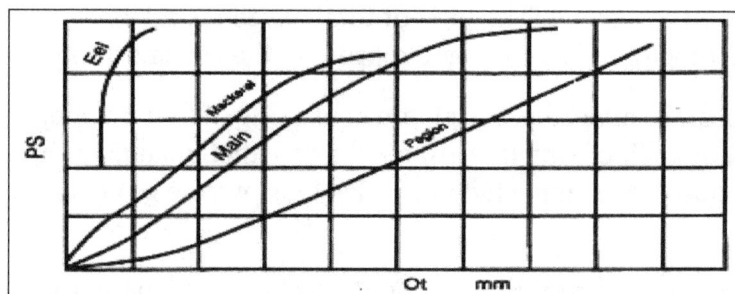

Graph showing comparative haemoglobin saturation curves as
a function of oxygen. PS, percentage saturation; ot, oxygen tension.

Bohr's Effect

It is the phenomenon during which if the partial pressure of CO_2 (PCO_2) increases, the higher tension is needed to reach TI, and Tu is decreased proportionally. The Bohr's effect is progressive in the fishes as compared to other vertebrates, and facilitates the unloading of oxygen to the tissue cells which contain comparatively high CO_2 tension.

Respiratory pigment consisting of polypeptide chain of amino acids whose carboxy and alfa amino groups are protonated accordingly to oxygenation and binding CO_2 to these group. The Bohr's effects differ according to species. The mackerel (Scomber scombrus) does not encounter low levels of oxygen and lives under uniform but low CO_2 tensions. They are found in high seas, whose blood is greatly affected by slight change in CO_2.

However, some fishes living in stagnant water such as Cyprinus carpio and bullhead catfishes (Ictalurus) have blood insusceptible to change in CO_2 concentrations. When CO_2 tension increases it results into the formation of H_2CO_3 in blood which readily dissociates into HCO_3^- and H^+.

An increase in hydrogen (H^+) causes lowering of pH and affects oxygen carrying capacity of haemoglobin. When pH falls from 9 to 6, it results the change in the shape of the oxygen dissociation curve of tuna haemoglobin, due to loss of co-operativity of the haem groups.

Effect of Temperature on Oxygen Dissociation Curve

When carbon dioxide tension is high and even with high partial pressure of oxygen as much as 100 atmospheres the blood does not saturate completely.

In this so-called root effect when CO_2 tension remains high with high partial pressure of oxygen, an increase in temperature raises the partial pressure required to saturate the blood. But in some fishes the absolute oxygen carrying capacity of the blood is higher at low temperature.

The warming of the water leads to increase respiration because the tissues demand more oxygen at higher temperature than at lower temperature. At the areas of human habitat having thermal and sewage effluents, mortality of fishes may occur due to asphyxiation caused by absolute lower lethal concentration of dissolved oxygen.

Oxygen Consumption

The rate of oxygen consumption is a measure of their metabolism, which can be regulated in several ways:

1. By the rate of oxidative metabolism.
2. By the flow of water over the gills causing diffusion gradient across the gill.
3. By the surface of the gill supplied by the blood.
4. By the area of gases exchange surface or the affinity of haemoglobin.

Generally the rate of oxygen consumption increases with rise in temperature up to some critical value, beyond which deleterious effects are visible and the rate falls off rapidly. It is also observed that the rate of oxygen consumption is usually lower in larger individuals of the species. The age, activity, nutrition, disease, reproductive state and nervous and hormonal control of the animal also affect the oxygen consumption.

Iris

The iris is a thin partition between the anterior and posterior chambers. It projects over the anterior surface of the lens with its free edge forming the pupil and controls the amount of light that reaches the retina.

The iris of elasmobranch's eyes has muscular element, so they adjust the shape of pupil. Majority of fishes have fixed pupil, either circular or oval. Their iris has no muscles but some guanine and melanin are present.

Lens

The lens is firm, transparent and ball-like and composed of non-collagenous protein. The lens is covered by the lens capsule and is filled by lens substance. Between them is present the lens epithelium, which plays an important role in metabolic processes of the lens. The lens substance is composed of lens fibres which are arranged in flat hexagonal prism. The lens fibres are highly modified epithelial cells. The bony fishes have roughly spherical lens. However, in sharks and rays the lens is horizontally compressed. Sometimes the lens is pyriform (Anablepidae) to provide for aquatic and aerial vision. In deep sea fishes, the eyes are projected with large lenses. Fish lens has a very high effective index of refraction. It is not homogenous and has an actual refractive index from 1.53 at the centre to 1.33 near the periphery.

In fishes, accommodation is usually achieved by altering the position of lens rather than changing lenticular shape. The lens is moved with the help of muscular papillae present on the ciliary body and adjustment of the vision occurs. However, some sharks and rays (Elasmobranchii) accommodate by small changes in lens convexity.

In trouts (Salmoninae) the lens has two focal lengths,: one is for the light rays reflected from a distant object lying lateral to the fish and focus on the central retina. Other is for the light rays reflected from close object and focus on the posterior retina. Thus these fishes have capacity to focus simultaneously the distant and nearby objects.

Consequently, the distance between the lens and central retina remains unchanged, and the focus remains adjusted for distant vision. At the same time the retractor lentis muscle contracts and brings the lens closer to the posterior retina, which is concerned with frontal field of vision of the fish.

The small space in front of lens is filled with clear, saline aqueous humor. The main cavity of eyeball is filled by the transparent vitreous humour which is secreted by the ciliary body.

Retina

It is the most important and sensitive part of the eye.

The retina is composed of several layers which from the outer to inner are as follows:

- Melanin containing pigment epithelium,
- Layers of rods and cones,

- Outer limiting membrane,

- Outer nuclear layer,

- Outer plexiform layer,

- Inner nuclear layer,

- Inner plexiform layer,

- Ganglion and nerve fibre layer,

- Inner limiting membrane.

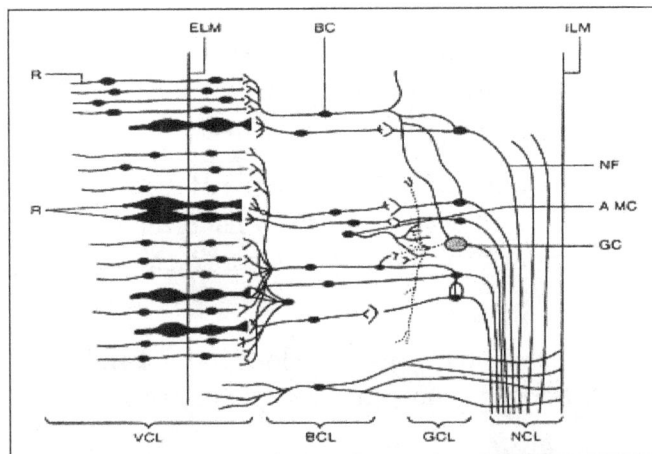

Diagram of nervous components of retina of Salmon. AMC, amacrine cell; BC, bipolar cell; BCL, bipolar cell layer, ELM, external limiting membrane; CC. ganglion cell; CLC ganglion cell layer; ILM, internal limiting membrane; NF, nerve fibre; NCL, nerve cell layer, R, rod; VCL, visual cell layer.

The retina has the following types of specialized nerve cells:

- Visual Cells: They are of two types, rod and cone cells. The rod cells are concerned to detect the intensity of light, while the cone cells distinguish wave length, i.e., colour.

- Horizontal Cells: They are present in the peripheral region of the inner nuclear layer. These cells give rise to some processes which extend horizontally near the outer nuclear layer and act as communicating lines between visual cells.

- Bipolar Cells: These nerve cells are found in innermost part of the retina. They are the largest neurons of the retina. The axons from these cells assemble and form optic nerve.

- Amacrine Cells: These cells are present between granular and inner plexiform layers and acts as horizontal lines of communication for visual stimuli.

The retina is well supplied with various capillaries which are of four types in the teleosts:

- Type A: having falciform processes.

- Type E: with vitreal vessels.

- Type F: with retinal vessels.

- Type G: unused instance.

Type A can be seen in the eye of rainbow trout which possesses falciform processes, containing the small blood vessels branches from the main vessel. The man vessel enters the eye from beneath the optic nerve papilla. The vessels are restricted to the falciform processes and are absent in the retina.

Type E blood vessels are present in the vitereo retinal boundary and are radially arranged from the centre, but they are absent in the retina. This type of intra-molecular vascularization is found in carp and eel. In Plecoglossus altivelis, embryonic fissure is present in the retina. Therefore, the vascularization patterns in the eye of Plecoglossus altivelis and carp is further classified into type E-a and type E-b respectively.

The epithelium of retina contains melanin and borders on the choroid coat. Melanin is photosensitive and in bright light it spreads and shades the sensitive rods; in weak light, it aggregates near the choroid border, thus photosensitive cells are fully exposed to the amount of light available.

Simultaneously the contractile myoid element in the base of the rods and cones move the cell tips in such a manner that rods are moved away from the lens in the bright light to be covered by the epithelium melanin, whereas in dark the rod myoids contracts and brings the cells towards the lumen of the eyeball.

Cones migrate in the opposite direction towards the lens in the bright light and towards the outer epithelium in dim light. Such photomechanical or retinomotor reaction is much pronounced in fishes than other vertebrates.

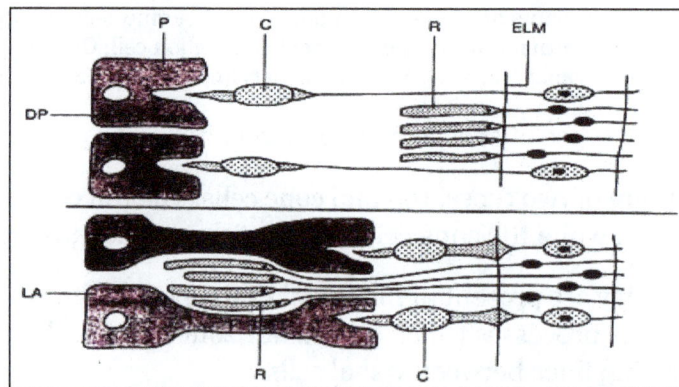

Diagram to show movement of rods and cones of retina of teleost
fish in light and darkness. C, cone; DP, darkness adapted; ELM,
external limiting membrane; LA, light adapted; P, pigment R. rod.

The relative number of rods and cones vary considerably in different species. In sight feeders which are active during the day, the cones are more in number than rods, and the revenue is found in many crepuscular species which are more active in twilight.

The retina of fishes contains two types of light sensitive pigments, rhodopsin and porphyropsin. Rhodopsin is purple coloured (marine Actinopterygii) while porphyropsin is rose-coloured (freshwater Actinopterygii). Migratory species like Atlantic salmon (Salmo salar), eels (Anguilla) and lamprey (Petromyzon marinus) possess both rhodopsin and porphyropsin.

More rhodopsin is present in the retina of lamprey migrating to the sea while porphyropsin predominates in the eye of spawners in freshwater. Rhodopsin and porphyropsin is synthesized by vitamin A in dark. In light there is found in the rods the yellow pigment retinine which can, however, be useful in resynthesizing the light sensitive purple or rose pigment.

The vision is photochemical process and involves reaction in the light sensitivity pigments of the rods and cones. But it is still not established that how these chemical changes become transformed into the electrical impulses that can be registered in retina nor how these ultimately converted into signals which is taken by optic nerve and travel to the brain.

Mouth

Fish mouths come in a variety of sizes, shapes, and orientations, each of which tells a great deal about what and where the fish eats, as well as something about its behavior. Predatory fish generally have the largest mouths, often sporting large, sharp teeth. Some species have mouths that can be extended, allowing the fish to lengthen its effective reach to catch tasty morsels of food as it swims. Other species have specialized mouthparts that allow them to rasp algae off rocks and branches. And additional varieties have mouths with teeth in the back, nearly in their throat. These pharyngeal teeth assist in grasping and swallowing prey.

Most fish mouths fall into one of three general types:

- Superior, or sometimes called supra-terminal, mouths are upturned.

- Terminal mouths point straight forward and are the most common mouth type.

- Inferior, or sub-terminal, mouths are turned downward. The inferior mouth type is often found in bottom-dwelling species, such as the catfish family.

Superior Mouth

The superior mouth is oriented upwards, and the lower jaw is longer than the upper jaw. Usually, fish with this type of mouth feed at the surface. They lie in wait for prey to appear above them, then strike suddenly from below.

Many species of fish with a superior mouth feed largely on insects, however, some may feed on other fish that swim near the surface. Some species with a superior mouth have an elongated lower jaw that functions much like a scoop.

Archers, half-beaks, and hatchetfish are all examples of species of aquarium fish that have a superior mouth.

Terminal Mouth

Terminal mouths are located in the middle of the head and point forward. Both jaws are the same length. More fish have this mouth type than any other. Fish having a terminal mouth are generally mid-water feeders; however, they can feed at any location. These species of fish are often omnivores, eating anything that is available. They typically feed on the move, either grabbing bits of food that they pass or preying on other fish that they chase down.

It is quite common for fish with a terminal mouth to also have a protrusible mouth which allows them to thrust the jaw forward when grabbing food. Most fish that feed on other fish have terminal mouths, which are often hinged to allow them to accommodate the action of snatching and swallowing another fish. They may also possess specialized teeth, and in some cases an additional jaw. Moray eels are one type of species that have a pharyngeal jaw placed well back in their throat.

Most barbs, cichlids, gouramis, and tetras have terminal mouths.

Inferior Mouth

Also called a sub-terminal or ventral mouth, the inferior mouth is turned downward. The lower jaw is shorter than the upper jaw, and the jaw will often be protrusible. Fish with inferior mouths are bottom feeders and often possess barbels that assist in locating food particles.

Most members of the catfish family have inferior jaws, and many of them also have a sucker mouth as well. The diet of fish with inferior mouths includes algae, invertebrates (such as snails), as well as detritus and any food that falls to the bottom.

Protrusible Mouth

A protrusible mouth allows a fish to extend its reach when attempting to snatch prey or food particles. This feature can be seen in all mouth types. Fish with a protrusible and hinged terminal mouth can create a vacuum when they open their mouths, thus sucking in their prey. Various species of fish may use a protrusible mouth on the fly while chasing down prey, while other species quietly lie in wait for prey to pass by, then rapidly extend the mouth to snatch the hapless victim.

Some species use this feature to engage in non-feeding activities. For example, kissing gourami uses its protrusible mouth to defend territory. Although it may appear to be kissing, it is a combative move to show its opponent who owns that space.

Other species, such as some members of the catfish family, use a protrusible mouth to stay in place by attaching to a rock or other stationary object.

Sucker Mouth

Sucker mouths are a common feature in fish with inferior mouths. Catfish, such as the popular pleco, use a sucker mouth to rasp algae off driftwood or rocks. Some species use a sucker moth to help those combat currents. By attaching itself to rock via its sucker mouth, it can stay where it wishes, even in a strong current.

These sucker mouths are also protrusible which allows the fish to extend its reach when sifting through the substrate for food particles. Sucker mouths can also be used when defending territory or quarreling with another fish.

Elongated Mouth

A greatly elongated snout is another kind of mouth adaptation. This type of mouth allows the fish to poke into small crevices and holes to find food. They may also use this mouth to dig through the substrate to reach buried food treasures. Some surface feeding fish also have an elongated mouth that allows them to scoop insects and food particles from the surface.

Freshwater species with elongated mouths include the halfbeaks, gars, and pencilfish. Saltwater species include the needlefish and the wrasse family.

Beak Mouth

The beak mouth is an interesting, but less common, mouth variation; it's also known as a rostrum. In this design, the mouth consists of two very hard pieces that are hinged and come together in a scissor-like fashion. This allows them to crush hard shells of invertebrates.

Pufferfish, both freshwater and saltwater varieties, possess a beak type mouth. Saltwater parrot-fish, octopus, and squid also possess a beak.

Lateral Line

The lateral line is a sensory system in fish and amphibians. It is made up of mechanoreceptors called neuromasts which are sensitive to water movement. The lateral line system has an important role in the detection of stationary objects, navigation, prey detection, capture and in swimming in schools.

The receptor organ of the lateral line system is the neuromast. There are two types of neuromasts, canal neuromasts which are located in the intradermal canals, and the superficial neuromasts which are located in the intraepidermal canals. Canal neuromasts are able to detect water flow acceleration, while superficial or free neuromasts can detect velocity.

In some species like the American paddlefish (Polyodon spathula), the lateral line system has evolved into an electrosensory system. This was accomplished by the specialization of hair cell receptors. These hair cell receptors in the lateral line system resemble the sensory hairs of insects. This may suggest that both derive from a common ancestral mechanosensory organ.

Structure of the Lateral Line System

Organization of the Lateral Line

The lateral line consists of a row of small pores which lead into the underlying lateral line canal. In the head, the lateral line canal is separated into three canals, one passes forward and above the eye, another forward and below the eye and the other downward and below the jaw. These three canals have numerous pores and together with the lateral line canal, make the lateral line system.

Epidermal structures called neuromasts form the peripheral area of the lateral line. Neuromasts consist of two types of cells, hair cells and supporting cells. Hair cells have an epidermal origin and each hair cell has one high kynocyle (5-10 Î¼m) and 30 to 150 short stereocilia (2-3 Î¼m). The number of hair cells in each neuromast depends on its size, and they can range from dozens to thousands. Hair cells can be oriented in two opposite directions with each hair cell surrounded by supporting cells. At the basal part of each hair cell, there are synaptic contacts with afferent and efferent nerve fibers. Afferent fibers, transmit signals to the neural centres of the lateral line and expand at the neuromast base. The regulation of hair cells is achieved by the action of efferent fibers.

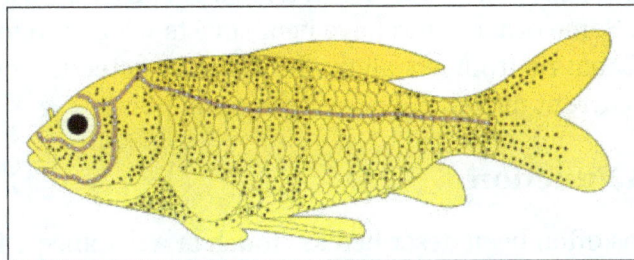

Diagram of the lateral line system.

In figure, the lateral line canal is divided into 3 stems, one passes forward and above the eye, another forward and below the eye and the other downward and below the jaw. Black dots represent the location of the neuromasts on the skin surface. White dots on the brown line show the positions of the neuromasts in sub-epidermal lateral line canals.

Stereocilia and kinocilium of hair cells are immersed into a cupula and are located above the surface of the sensory epithelium. The cupula is created by a gel-like media, which is secreted by non-receptor cells of the neuromast. There are two types of neuromasts, superficial or free neuromasts and canal neuromasts. Superficial neuromasts are located at the surface of the body and are affected by the environment. Superficial neuromasts are categorized into primary or paedomorphic neuromasts and secondary or neomorphic neuromasts. Canal neuromasts are primary neuromasts. These are found inside epidermal or bony canals and are located on the head or body of the fish.

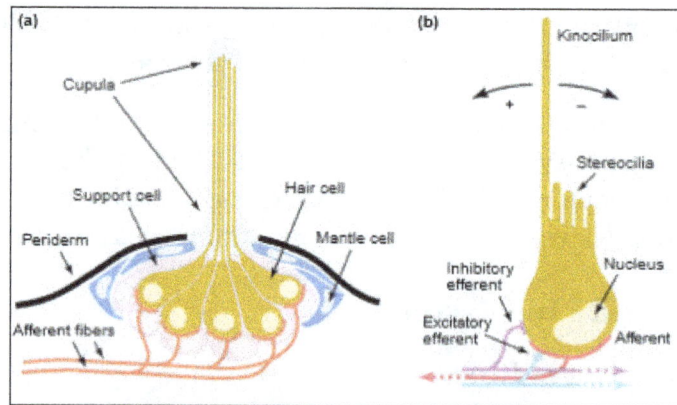

Lateral line of fish. (a) The figure shows the basic structure of
neuromasts and all its components. (b) Hair cell, depicting
the innervation of afferent and efferent fibers.

Superficial and Canal Neuromasts

Superficial neuromasts are small and can be found in lampreys, teleost fishes and in some bony
fishes. Superficial neuromasts are located on the head and the body and in some fish in the caudal
fin.

They have a cylindrical cupula and a round base with a diameter that can seldom reach 100 Î¼km.
The number of hair cells is small, from several dozens to several hundred.

In canal neuromasts, the sensory area is situated at the bottom of the canal below the skin. Canal
neuromasts have a large range in size, shape and orientation within the canal. Some species have
narrow canals and the neuromast can be found in a local constriction with the long axis running
parallel to the canal axis. Some other fishes have neuromasts which are found in wide canals and
have a different shape. Canal neuromasts allow the efficient detection of pressure differentials,
which are created by the current movement across the canal pores.

Lateral Line System Function

The lateral line system has often been described as "touch at a distance". This is due to the lateral
line function being similar to the senses of touch and hearing. The earliest hypothesis about the
function of the lateral line was that it secretes mucus to cover the body. Several years later, it was
determined that the lateral line is used to detect water current and stimuli from moving objects.

Fish can sense water movements ranging from large-scale currents to small disturbances caused by
plankton. This is due to the superficial neuromasts which are able to respond to very weak water
currents, with speeds from 0.03 mm/s and higher. Canal neuromasts can respond to current speeds
from 0.3 to 20 mm/s. The lateral line has functions in schooling, prey detection, spawning, rheotax-
is (which is a form of taxis when fish face an ongoing current), courtship and station holding.

It is thought that the lateral line system can create hydrodynamic images of the surrounding area.
This can be achieved by detecting moving and stationary objects in active and passive ways. Active
hydrodynamic imaging is similar to the echolocation of objects that is observed in dolphins. Here,
fish produce a flow field around their body, which helps them in detecting distortions in their flow

field. This is observed in blind cavefishes, which rely on this mechanism to explore their surroundings. For example, they are able to differentiate between structures that differ by even 1 mm.

Passive hydrodynamic imaging can be carried out for moving and stationary bodies. This is achieved by detecting currents that are generated by other moving bodies such as other fish or the movement of stationary objects such as rocks in a stream.

Lateral Line Information Processing

Lateral line information is processed in all regions of the brain. The information is provided by afferent nerve fibres and is sent to the brain via the lateral line nerves that enter the ipsilateral brainstem and terminate in the medial octavolateralis nucleus (MON). Main primary lateral line projections reach the ipsilateral cerebellar granular eminence while the second order of projections from the medial octavolateralis nucleus terminate in the lateral compartment of the torus semicircularis and in the deep layers of the optic tectum. The final pathway for information processing is the relay of information from the midbrain to different diencephalic nuclei.

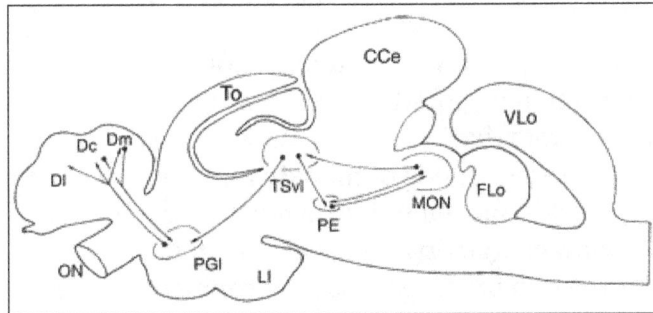

This figure depicts the pathway of information processing. MON represents the medial octavolateralis nucleus, CCe represents the Corpus cerebelli, Ll is the hypothalamic inferior lobe, Flo is the facial lobe, ON is the olfactory lobe, PGl represents the lateral preglomerular nucleus, PE is the pre eminential nucleus, TSvl is the Ventro lateral nucleus of torus semicircularis.

Lateral Line Modifications

The lateral line system of elasmobranchs is different to that of teleost fish. Elasmobranchs have superficial neuromasts and two morphological classes of sub-epidermal canals. Elasmobranch canals have skin pores that allow direct contact with the surrounding water. They may also have absent skin pores which prevent the contact of canal fluid with the external environment. In teleost fish, hydrodynamic pressure differences at the skin pores cause fluid motion. This results in pored canal neuromasts being able to cipher the acceleration of external water flow near the skin, and induce behaviours such as hydrodynamic imaging, detection of prey and schooling. In elasmobranch fishes, other than prey detection the function of the lateral line pores and their neurophysical response is not yet known.

Sharks and batoids have non-pored canals which are located on the ventral body surface, rostrum and around the mouth. The absence of skin pores demonstrates that localized weak hydrodynamic flow which causes pressure differences will not produce canal fluid motion directly, as it occurs in the pored canal systems.

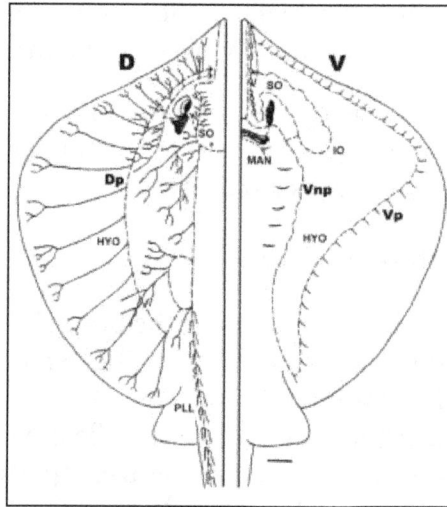

Lateral line canal system on the dorsal (D) and ventral (V) surface of the Atlantic stingray,
Dasyatis Sabina. Solid lines indicate neuromast-free tubules which terminate in pores.
The other lines indicate canal sections which consist of innervated neuromasts.

A hypothesis was developed to explain the function of non-pored canals in elasmobranch fishes. The hypothesis explains that the non-pored canals of stingrays which are located on the ventral surface function as tactile receptors that aids in the localization and capture of small benthic prey. The hypothesis explains that direct coupling of the skin and canal fluid should result in an increase in sensitivity to the velocity of skin movement, which would mean that primary afferents that innervate neuromasts would show characteristics consistent with detectors of velocity. The hypothesis also states that without direction to the external environment, non-pored canals will have lower sensitivity to water motion in comparison to tactile stimulation.

Luminous Organs

A number of fishes especially marine species are known to produce characteristic light through their special organs called luminous organs. These organs are commonly found in fishes living in deep-sea where the sunlight ceases to enter. The luminous organs are absent in freshwater fishes.

The most important function of bioluminescence is to illuminate surroundings for the purpose of camouflage, schooling and for recognition of movement of predators in the water. The luminous organs or photophores are special gland cells of the epidermis. Their distribution on the body type and adaptive value may vary in different species of fishes.

Structure of Luminous Organs

On the basis of anatomy of photophores they may be categorized in two types:

Simple Photophore

They are small in size, about 0.1 to 0.34 mm in width. It consists of light generating cells called as photocytes. Simple type may be provided with or without mantle of pigment. The lenses are formed by grouping of cells known as lenticular cells.

The distal part of photocyte is provided with acidophilic granules. A layer of melanophres surrounds the photophore. Simple type of photophores is present in sharks. In Stomias the luminous organs are logged in gelatinous corium of the epidermis.

Compound Photophore

This type of photophores consists of additional structures like reflectors, pigmented mantle and sub-ocular organs. The latter one is a large organ deeply embedded in dermal tissue. The photocytes are arranged in the form of cords and bands.

Photogenic tissue, pigment and reflector layers are provided with nerves and blood vessels. The photogenic tissues are found in the centre of the photophore and consist of two types of glandular cells.

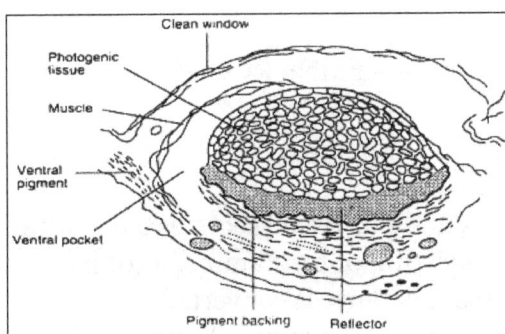

T.S. through a sub-ocular light organ of Astronesthes richardsoni.

The mechanism of light production is peculiar in fishes and takes place the special sets of muscles present around the photocytes. When these muscles contract, they pulls the outer surface of photophore downwards causing brighter surface to be concealed.

In contrast the relaxation of these muscles exposes bright surface of the photophores. In some species, movement of pigmented layer carries out concealing and rotating of photophores.

Types of Luminous Organs

On the basis of source of illumination it may be classified as follows:

Extra Cellular Luminescence

Light may be generated by luminous secretion from the glandular tissues. Extra cellular luminescent organs are found in a very limited species of fishes. Certain fishes like rat tails and searssids emit light by secreting extra cellular slime. Rat tails possess special glands near its anus, which secretes slime of sufficient luminosity.

Intracellular Luminescence

In this type the light is produced within the glandular cell or intrinsic photocyte. These luminous organs developed from the epidermis.

Fishes ornamented with this type of luminous organs belong largely to the family of teleosts, i.e., Sternoptychidae (hatchet fish), Myctophidae (lantern fish), Halosauridae (Halosaurid eel),

Stomiatidae (scaly dragon fishes), Brotulidae (Brotulus), Lophiidae (anglerfish) and Zoarcidae (eel pouts).

Bacterial Luminescence

In this type, symbiotic bacteria present in the photophore or luminous cell discharge light. Many different species are recognized particularly the genus photo-bacterium and achromobacterium have been isolated and grown in cultures. They are common on dead fish or spoiling meat.

The biochemical step in bacterial luminescence is linked to the electron transport chain of oxidative phosphorylation, in which flavin mononucleotide ($FMNH_2$) from the electron transport chain reacts with an aldehyde (RCHO) to form a complex (luciferin) that is oxidized to an acid (RCOOH) with emission of light.

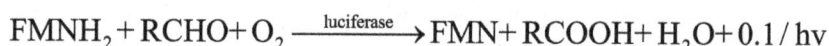

$$FMNH_2 + RCHO + O_2 \xrightarrow{\text{luciferase}} FMN + RCOOH + H_2O + 0.1/hv$$

Chemical Luminescence

It has been established that the glandular tissue secretes a chemical substance called as luciferin, which is an indole derivative consisting of tryptamine, arginine and isoleucine. Under the influence of the enzyme luciferase, this substance is converted into oxy-luciferin and emits blue or blue-green light. Apogon, the Parapriacanthus is known to possess luminous glands containing crude form of luciferin and luciferase.

Control of Luminous Organs

The function of light producing organs is controlled by the nervous or endocrine system.

1. Nervous Control: Several workers have reported that light production by the luminous organs is controlled by the nervous system, probably by the peripheral sympathetic system. The nerves innervate the phagocytes. The efferent nerves enter the photogenic cells and activate them.

2. Hormonal Control: It has been reported that some fishes have hormonal control on the photophores. Endocrine gland like supra renal activate them. Adrenalin or noradrenalin is known to control light emission from the photophores.

3. Mechanical Control: The muscles present beneath the photophores contract and rotate the photophores in such a way that they get concealed. Thus fish is prevented from illumination specially when in danger.

In Photoblepharon palpebratus the ventral part of luminous organ has a fold of black tissue. This fold can be drawn over the photophores and conceal the light. In some fishes the light production is also supposed to be influenced by the movement of pigment in the chromatophores.

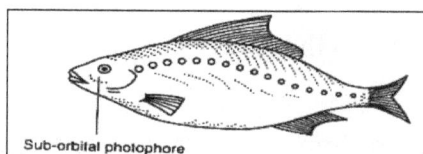

Light producing organs of Photoblepharon.

Biological Significance of Luminous Organs

This is useful in variety of ways in marine fishes specially in deep-sea fishes.

Illuminates Surroundings

Some fishes utilize their luminous organs to illuminate their surroundings in the event of dimness. Thus they become able to search their prey in the dark waters. Some species (stomiatoid) are able to emit beam of light from the specially designed luminous cheek organ to catch the small creatures like planktons. The cheek organs of Anamalops produce light like a torch.

As Defensive Device

Many fishes produce sudden flash of light from their luminous organs, which helps in diverting the attention of their predators. The emission of light facilitates an escape of fish by puzzling the enemy. Alepocephalidae produce a glowing spark, which confuses the predator for a spur of moment, and help the fish to escape.

However, some fishes use luminous organs to enable them inconspicuous. In doing so they illuminate their ventral surface that makes them inconspicuous against lighted background above.

As a Warning Signal

A number of fishes use its luminous organ to warn the predators. For instance, the midshipman Porichthys that possesses, a toxic sign, flashes light when it is attacked by a predatory fish and avoids the danger.

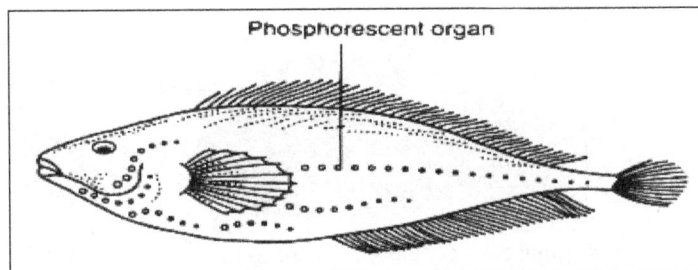

Hight producing organs of Porychthyes.

Recognizing Own Species

Every species has a unique arrangement and distribution of photophores on their body, which help the fish to recognize species of same type and thus help in schooling behaviour. The luminous organs are also helpful in recognizing the mates for courtship, as the light organs may be different both in male and female.

Male lantern-fish has one or many photophores present above but both in the female possesses it below the caudal peduncle. In some species the size of luminous organ is different in both sexes. For example in many species of melanostommiatidae, the postorbital luminous organ is larger in the male and smaller in the female.

Internal Fish Anatomy

Spinal Cord

All vertebrate animals have spinal cords. Phylogenetically, it is the oldest part of the central nervous system (CNS). In contrast to the more recently evolved cerebral and cerebellar hemispheres of the brain, the cell bodies and dendrites of spinal neurons (gray matter) lie inside the cord while the nerve fibers (axons) that interconnect them (white matter) run along the outside. As these axons ascend and descend along the body, the white matter occupies more space as the cord approaches the head. All movements of the body below the head are controlled by the spinal cord and injuries to it produce devastating losses of function.

The basic structure of spinal cord in fishes resembles to that of higher vertebrates. It runs dorsal and lengthwise of the fish body in the neural canal of the vertebral column. In transverse section of spinal cord two regions are clearly distinct, i.e. central and peripheral. The central region is composed of numerous nerve cells, hence looks grey and called grey matter, which roughly resembles a letter 'Y' or 'X'.

The peripheral region, which surrounds the grey matter, contains large number of militated nerve fibres and is known as white matter. The centre of grey matter is occupied by a central canal, which is filled with cerebrospinal fluid (CSF). The cerebrospinal fluid is secreted by the brain. The central canal is lined with ependymal cells.

The central canal is formed by extension of brain ventricles. The grey matter has an anterior, posterior and lateral columns or horns from where the spinal nerve fibres enter or leave the spinal cord. The anterior or ventral roots arise from the anterior columns to innervate somatic muscles or visceral organs.

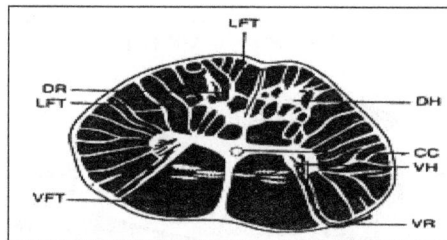

In figure, section of spinal cord of shark CC, central canal; DFT, dorsal fibre tract; DH, dorsal horn; DR, dorsal root of spinal nerve; LET, lateral fibre tract; VFT, ventral fibre tract; VH, ventral horn; VR, ventral root fibres of spinal nerve.

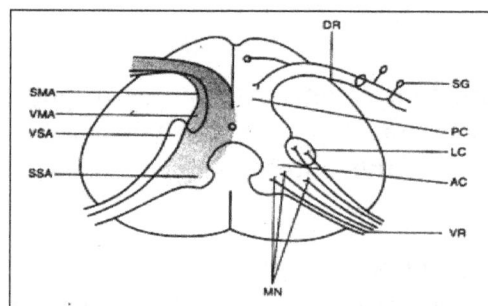

Diagram of spinal nerve showing sensory and motor areas. AC, anterior column, DR, dorsal root; LC, lateral column; MN, motor neuron; PC, postesior colurtm; SC, spinal ganglion; SMA, somatic motor area; SSA, somatic sensory area; VR, ventral root; VSA, visceral sensory area.

Those nerves which innervate visceral effectors comprise a pair of sympathetic ganglion present along their pathways. However, the peripheral stimuli transmit through another ganglion found in the posterior root of the posterior column of spinal cord. Then these stimuli are conveyed to nerve cells in the central nervous system.

The fish spinal cords possess two types of cells in the grey matter. One kind of cell generally found in the dorsal region and plays an important role in controlling the movements of trunk musculature.

The types of cells are present in the central region of the grey matter. The fibres originating from these cells innervate certain effectors such as the fins. The spinal nerve fibres are arranged into two roots, one is the posterior sensory root and another is anterior motor root.

Brain

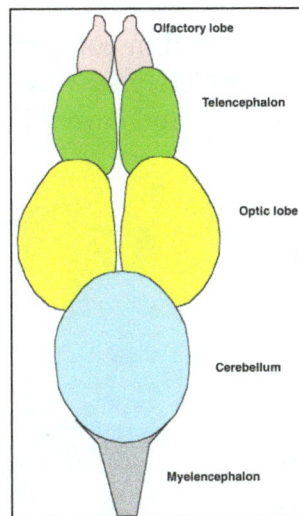

The brain of a rainbow trout. Fish brains are divided into a
number of different regions each specialized to carry
out different neuronal processing functions.

Brain is the mass of nerve tissue in the anterior end of an organism. Fishes have a centralized nervous system with a brain located in a distinct head region. Compared to other vertebrates, the brain is relatively small. On average it is 1/15 the size of the brain of a bird or mammal of equal body size. But relative to almost all invertebrates, it is a highly advanced organ that is capable of processing complex sensory information. In fact, some fish, such as sharks, have large brains that are comparable to mammals. The fish brain is divided into several different regions:

- Olfactory lobes

- Telencephalon

- Diencephalon

- Pineal body

- Optic lobes

- Cerebellum

- Metencephalon

- Myelencephalon

Each region is specialized to carry out different functions. For example, the olfactory lobes are involved in processing sensory input from the nostrils, whereas the metencephalon primarily carries out nervous system processing related to swimming and maintaining balance. A diagram of the brain of a rainbow trout, showing several of these regions, is shown in figure.

Development of Brain in Fishes

During the development the brain distinguishes as an anterior enlargement of the spinal cord. The brain is the regulating centre for all the receptors. The brain is enclosed in a cranium, which is cartilaginous in elasmobranchs and bony in teleost.

Generally, the brain of fishes is relatively small in comparison to their body size and hence the brain does not occupy the cranial cavity completely, leaving small gap, which is filled with a sort of gelatinous matrix. The brain is soft and white and covered by extensive network of blood vessels called choroid plexi.

Divisions of Brain in Fishes

The brain has three main divisions i.e.:

- Prosencephalon or forebrain

- Mesencephalon or midbrain

- Rhombencephalon or hindbrain

The general organization in the major groups of fishes is not dissimilar, although considerable differences occur in the form and enlargement of different parts of the brain.

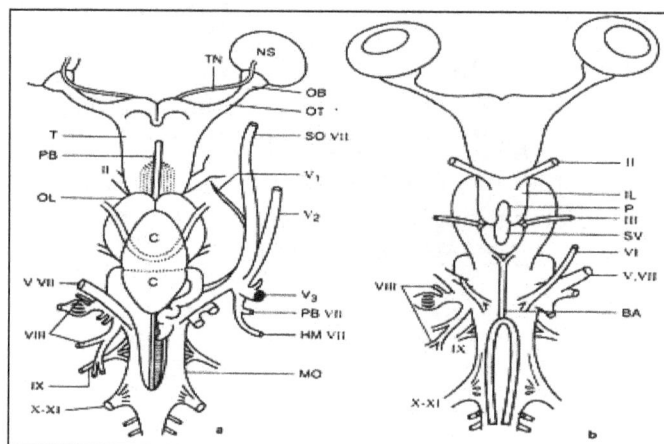

In figure, Brain and cranial nerve view of Squatus (a) dorsal view. (b) ventral view. BA, basilar artery; C, cerebellum; HM, hyomendfbular; NS, nasal sac; IL. Ulterior lobe; MO, medulla oblongata; OB, olfactory bulb. OL, optic lobe; OT, olfactory tract; P, pituitary; PB, palatine branch; SO, superficial ophthalmic SV, Saccus vasculosus; T, telencephalon; TN, terminal nerve; II-IX-cnutial nines.

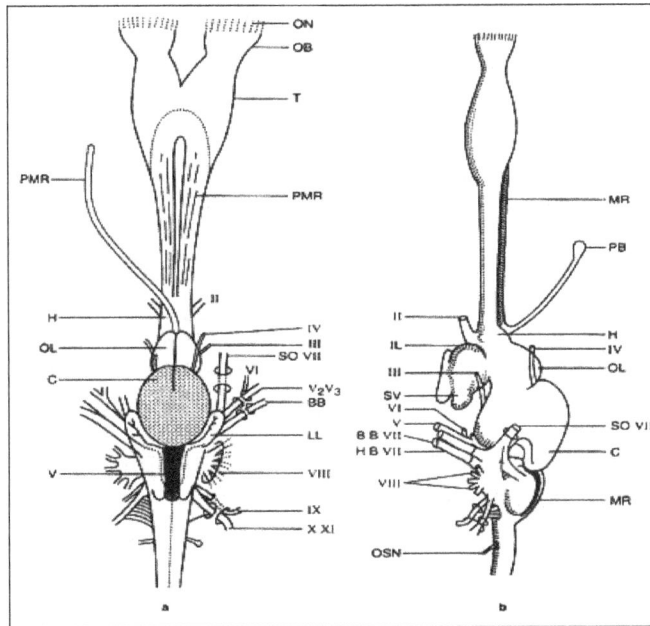

In figure, Brain and cranial roots of Hydronagus. (a) Dorsal view. (b) lateral view. BB, buccal branch; C, cerebellum; PV, fourth ventricle; H. hebenula; LL, lateral-line lobe; MR, membranous roof; Off, olfactory bulb; ON, olfactory nerve; OL, optic lobe; OSN, Occipetostinal nerve; PB, pineal body; PMR, paraphyals in membranous roof; SO, superficial opthalmic VII. SV, saran vasculosus; T, telencephalon. II-XI are cranial nerve.

The homologies of brain parts are summarized.

Embryonic parts			Definite brain parts	Brain cavities
Prosencephalon	1. Telencephalon		Forebrain (Cerebral hemisphere)	Paired lateral ventricals
Embryonic parts				
	2. Diencephalon		Between brain	Third ventricle
Mesencephalon	Mesencephalon		Midbrain	
Rhombencephalon	Metencephalon		Cerebellum	Metacoel
	Mylencephalon		Medulla oblongata	Fourth ventricle

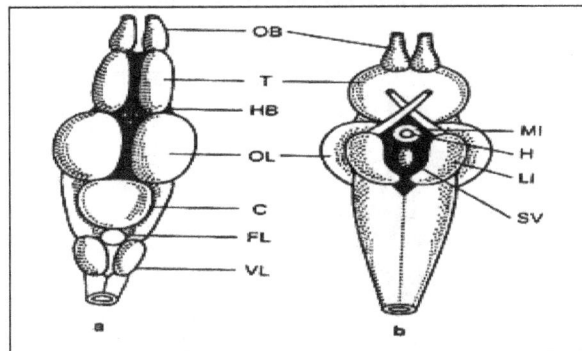

Diagram of brain of fishes. (a) Brain of Puntius tido (dorsal view). (b) brain of S. diaconthus.
C, cerebellum; FL, facial lobe; H, hypophysis; Hb, habenula; L3, inferior lobe; Ml, median infundibulum; OB, olfactory bulb; OL, optic lobe; SV, saccus vasculosus; T. telencephalon; VL, vagal lobe.

In above figure, (a) Cat shark. (b) Smooth dogfish. (c) Sand shark. C, cerebellum, F, folds; OT, optic tectum; SC. spinal cord; TO, transverse groove; TS towards snout.

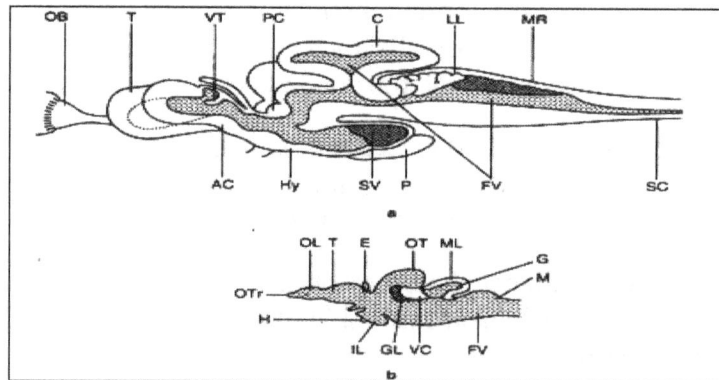

Lateral view of brain of fishes.

In figure, (a) Squalus. (b) puntius tido. AC, anterior commissure; C, cerebellum; E, epiphysis; FV, fourth ventricle; CL, granular layer; Hy, hypophysis; IL, inferior lobe; LL, lateral lobe; M, medulla; MR, membranous roof; OB, olfactory bulb; OL, olfactory lobe; OT, optic tectum; OTr, olfactory tract; P, pituitary; PC, posterior comminute; SC, spinal cord; SV, seats vascutosis; T, telencephalon; VC, vulvula certbelli; VT; velum transversurn.

Regarding the function of pineal body, two theories have been put forward. According to the first hypothesis, the pineal body has sensory roles while the second hypothesis states its secretory function. In sensory roles it acts as a baro or chemoreceptor for cerebrospinal fluid or helps in facilitating the olfactory response to sex hormones. The secretory functions involve the external secretion related to the chemical composition of the cerebrospinal fluid or the metabolism of brain tissue. It is also expected to be an endocrine gland with internal secretions.

The thalamus consists of lateral walls of the diencephalon. The thalamus acts as a relay centre for transmitting olfactory and strait body impulses to thalamomedullar and thalamospinal tract. The ventral part of thalamus bears geniculate nuclei or ganglia. This ganglion is well developed in sharks and is known as geniculate lobes.

The hypothalamus constitutes the floor of the diencephalon. It is the largest and the most important part of diencephalon. Anterior part of the hypothalamus contains preoptic area. However, from the ventral part of hypothalamus projects a pouch-like down growth, known as the infundibulum.

The tip of infundibulum bears hypophysis or pituitary gland. There are two important nuclei centres present in the hypothalamus, one is nucleus preopticus and the other is nucleus lateralis tuberis. Both are made up of nerve cells of secretory nature. In Osteichthyes, the lower part of the hypothalamus extends laterally which are known as inferior lobe.

The saccus vasculosus is highly vascular protrusion of the ventral wall of the diencephalon of fishes. It is regarded as benthic organ detecting changes in water pressure. It plays a secretory role also. The diencephalon acts as an important correlation centre for incoming and outgoing message concerned with internal homeostasis. The hypothalamus plays an important role in regulating the endocrine system through pituitary gland.

Mesencephalon or Midbrain

The mesencephalon or midbrain is relatively large. It is made up of the dorsal optic tectum and the ventral tegumentum. Fishes that feed by sight possess large optic lobe. The optic tectum appears as two optic lobes from the dorsal side.

The ventricle communicates the mesencephalon as narrow cavity, called aquaductus mesencephali or aqueduct of Sylvius. There is considerable variation in the histological structure of the optic tectum.

The optic tectum is composed of six layers which are as follows:

1. The superficial layer, the stratum opticum which is innervated by optic nerve fibres.

2. Stratum fibrosum, griseum superficial, which is an important seat of visual sense and possess nerve fibres and nerve cells.

3. Stratum griseum centrale consists of nerve cells of efferent nerves.

4. Stratum album centrale, it consists of nerve fibres of efferent nerves and is associated with oculomotor and thalamus.

5. Stratum griseum periventricular comprising nerve cells which is connected with the fibres of the periventricular system.

6. Stratum fibrosum periventricular, lies in front of aquaductus mesencephalic and possess numerous nerve fibres.

Because of these multi-layered arrangements of the nerve cells in optic tectum, it is often considered to be homologous with the cerebral cortex of mammals.

The mid-ventral part of the optic tectum contains a protuberance, called torus longitudinalis, which helps in integration between the sense of equilibrium and the sense of vision. There is a strong evidence for optic tectum as eye body coordinating centre and compensate the absence of a true optic chiasma of bony fishes.

Electric stimulation of tegumentum results in uncoordinated locomotory response. Optic tectum is an important correlation centre for various types of impulses of the brain.

Rhombencephalon or Hindbrain

The hindbrain consists of cerebellum or metencephalon and medulla oblongata or mylencephalon.

a. Cerebellum or Metencephalon:

The cerebellum develops as prominent dorsal outgrowths from the medulla. Its anterior part projects forwardly into the cavity of optic lobes and is known as valvula cerebelli which is composed of cortex and medulla. The cortex is made up of three distinct layers comprising differently shaped nerve cells.

1. The molecular or chief receptive layer comprising Purkinje cell dendrites and glia cells.

2. The granular cell layer consists of small granular cells and also receives the Purkinje cell axons. Fishes having well developed gustatory (carp) and lateral line (catfish) system have a more developed vulvula cerebelli.

They are the largest component of the brain and project forwardly beyond the forebrain in the mormyrids. Since mormyrids generate and respond to weak electric currents, it is considered that their enlarged cerebellum is involved in the reception of electrical impulses.

Elasmobranchs shows an increase in size of cerebellum as compared to the Cyclostomes. The dog-fish (small shark) possess simple bilobed cerebellum. However, the mackerel sharks which are large and swift have the larger cerebellum occupying much part of the brain. In these fishes the cerebellum becomes convoluted to increase surface area.

The electric current discharging fishes such as mormyrids and the electrical catfish (Malapterurus) contains very well developed cerebellum dealing with electrical impulses.

The cavity of the cerebellum is known as metacoel, which is prominent in sharks and rays (Elasmo-branchs) and is completely disappeared in cerebellum of higher bony fishes. The main function of the cerebellum is to control swimming equilibration, maintenance and co-ordination of muscular tonus, and orientation in space.

b. Medulla Oblongata or Mylencephalon or Brain Stem:

The medulla oblongata or mylencephalon is the posterior most part of the brain, which can be distinguished from the spinal cord. The medulla is divisible into columns of nerve fibres based on the types of information transmitted. Thus there are visceral and somatic sensory and visceral and somatic motor columns.

The medulla contains nuclei of cranial nerves from III to X, arranged anteroposteriorly. The different parts of medulla are enlarged with the development of various senses. The medulla contains a cavity inside, known as the fourth ventricle. Some fishes such as Culupea and Mugil possess prominent parried swellings called 'cristae cerebelli' present on the anterolateral boundary of the fourth ventricle.

These cristae are associated with schooling behaviour of these fishes. In goldfish prominent vagal lobes are present behind the cristae cerebelli, from which IX and X cranial nerves arise. This fish also possesses a palatal organ found in the roof of mouth, helping in testing the food by taste and touch.

Another prominent structure, the facial lobe or tuberculum. impar found behind the cerebellum in Cyprinus carpio. In the facial lobes the gustatory and tactile impulses are corrected with visceral sensory ones as it arises from the fusion of root elements of the facial (VII) and (X).

The medullas of higher bony fishes (Actinopterygii) have a pair of large neurons called the giant cells of Mauthner. These cells are present at the level of cranial nerve VIII. In Anguilla and Mola the Mauthner cells are absent whereas less prominent in bottom dwellers such as gobies and scorpion fishes. These cells are motor coordinator for transmitting multiple sensory impulses mainly from lateral-line centres to the caudal and body swimming muscles.

The medulla is sensory and sensory coordinating centres and a relaying area between the remaining brain and spinal cord. Medulla also has certain centres that control some somatic and visceral functions. Among bony fishes these functions includes respiratory, paling of body colour and osmoregulatory.

Swim Bladder

In most of the fishes a characteristic saclike structure is present between the gut and the kidneys. This structure is called by various names, viz., swim-bladder, or gas-bladder, or air-bladder.

The swim-bladder occupies- the same position as the lungs of higher vertebrates and is regarded as homologous to the lungs. It differs from the lungs of higher forms mainly in origin and blood supply.

The swim bladder arises from the dorsal wall of the gut and gets the blood supply usually from the dorsal aorta, while the vertebrate lung originates from the ventral wall of the pharynx and receives blood from the sixth aortic arch.

The swim-bladder is present in almost all the bony fishes and functions usually as a hydrostatic organ. Starting as a very insignificant cellular extension from the gut, the swim-bladder in fishes leads the whole group through an evolutionary channel.

Development of Swim-bladder

Opinions differ as regards the development of swim bladder in fishes. In teleosts, it originates as an unpaired dorsal or dorsolateral diverticulum of the oesophagus. It starts as a small pouch budded off from the oesophagus. The diverticulum with an opening in the oesophagus becomes subsequently divided into two halves.

Of these two, the left one often atrophies except in a few primitive forms. The right half becomes well-developed and takes a median position. In dipnoans and Polypteridae, the swim-bladder is modified into the 'lungs' and originates as the down-growths from the floor of the pharynx.

These out-growths have been rotated around the right side of the alimentary canal to occupy the dorsal position. As a consequence of shifting of the position, the original right 'lung' becomes the left one. Spengel advocates the view that the swim-bladder in fishes originates from the posterior pair of the gill-pouches, but definite embryological evidence in support of this idea is lacking.

Basic Structure of Swim-bladder

The swim-bladder in fishes varies greatly in structure, size and shape.

1. It is essentially a tough sac-like structure with an overlying capillary network.

2. Beneath the capillary system there is a connective tissue layer called tunica externa.

3. Below this layer lies the tunica interna consisting primarily of smooth muscle fibres and epithelial gas-gland.

4. The swim bladder lies below the kidneys, between the gonads and above the gut.

5. The connection with the oesophagus may be retained throughout life or may be lost in the adult.

Gas Composition of Swim-bladder

the gas secreted by the swim-bladder is mostly oxygen. Nitrogen and little quantity of carbon-dioxide are also present. Generally the gas composition varies in different species. In salmonids, the maximum amount of gas in the swim-bladder is Nitrogen. Again in many species the composition includes mostly a mixture of oxygen and carbon dioxide.

Types of Swim-bladder

Depending on the presence of the duct (ductus pneumaticus) between the swim-bladder and the oesophagus, the swim-bladder in fishes can be divided into two broad categories: Physostomous.

Depending on the condition of the swim-bladder, the teleosts are classified by older taxonomists into two groups Physostomi and Physoclisti. A transitional condition is observed in eels.

Physostomous Condition

The swim-bladder develops from the oesophagus. When the ductus pneumaticus is present between the swim-bladder and the oesophagus, the swim-bladder is called physostomous type.

A vessel emerging from the coeliacomesenteric artery supplies the swim bladder and the blood from it is conveyed to the heart through a vein joining the hepatic portal vein. This condition is observed in bony ganoid fishes, the dipnoans and soft-rayed teleosts.

Physoclistous Condition

In this condition the ductus pneumaticus is either closed or atrophied. This type of swim bladder is observed in spiny-rayed fishes. In this type of swim-bladder, there lays an anteroventral secretory gas gland (containing retia mirabilia) and a posterodorsal gas absorbing region called the oval. The oval develops out of the degenerating ductus pneumaticus.

The rete mirabilis of the gas gland, the oval and the walls of the bladder are supplied by the coeliacomesenteric artery and also by arteries from the dorsal aorta. But the blood from the different parts of the swim bladder is returned by two routes.

The blood from the gas gland is returned to the heart by the hepatic portal vein, while from the rest of the bladder by the posterior cardinal veins. The bladder, specially the gas gland, gets the lateral branches from the vagus, while the oval is innervated by sympathetic nerves.

Transitional Condition

In Eel (Anguilla), a transitional condition between the physostomous and physoclistous type is present. The swim-bladder retains the ductus pneumaticus which becomes enlarged to form a separate chamber containing the oval. The gas glands are also present.

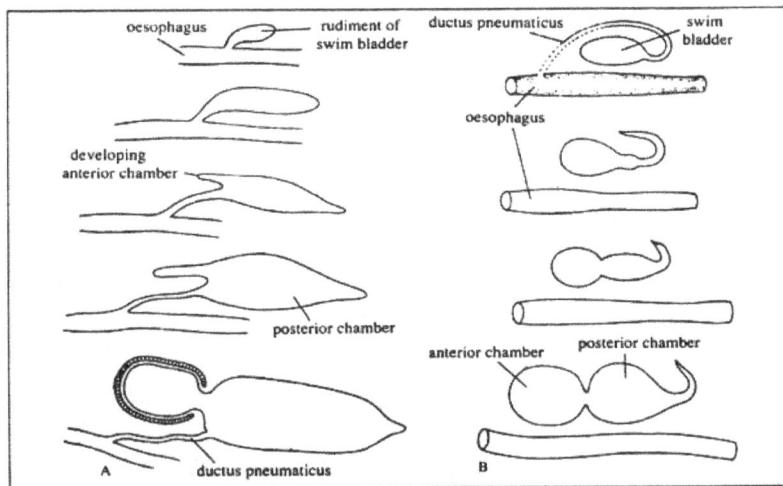

Showing the derivation of the swim-bladder of the fishes from the gut.
A. Stages of formation of physosto-mous type of swim bladder in Catostornus.
B. Stages of fotmation of the physoclistous type of swim bladder.

The swim-bladder is supplied with the blood through a branch from the coeliacomesenteric artery while the blood is returned to the heart by a vessel joining the post cardinal vein. The condition represents an intermediate stage when a physostomous condition is on the verge of transformation into the physoclistous state.

Modifications in Swim-bladder

In fishes a great diversity in size, shape and function of the swim-bladder is observed. In elasmobranchs, bottom dwelling and deep-sea teleosts the swim-bladder is absent in an adult but a transitory rudiment during development may be present.

In flat fishes (Pleuronectidae) swim-bladder is present in the early life when the animals maintain a vertical position. As they tip over one side and assume the lazy adulthood, the swim-bladder becomes atrophied.

In elasmobranchs, the swim-bladder is represented by the transitory rudiment in the embryonic stages. Miklucho-Maclay has observed a rudimentary dorsal diverticulum from the foregut in the embryos of Squalus, Mustelus ana Caleus. In many fishes, viz., Heptranchias, Scyllium, Squatina, Pristiurus, Carcharius and many Rays, small pits are recorded in the oesophageal wall.

Wassnezow has observed one to six similar oesophageal pits in Pristiurus, Torpedo and Trygon. These pits are located posterior to the fifth pouch. In sharks the swim-bladder is absent in adults, but a hint of a rudimentary swim-bladder is observed during embryonic development. But almost all the teleosts possess the swim-bladder and extreme modifications of the same are encountered because of adaptation to the different modes of living.

Modifications of Physostomous Condition

The typical physostomous pattern becomes modified in different fishes and the basic trends are:

1.   The formation of paired sacs.

2.   The gradual acquisition of two chambers— an anterior and a posterior.

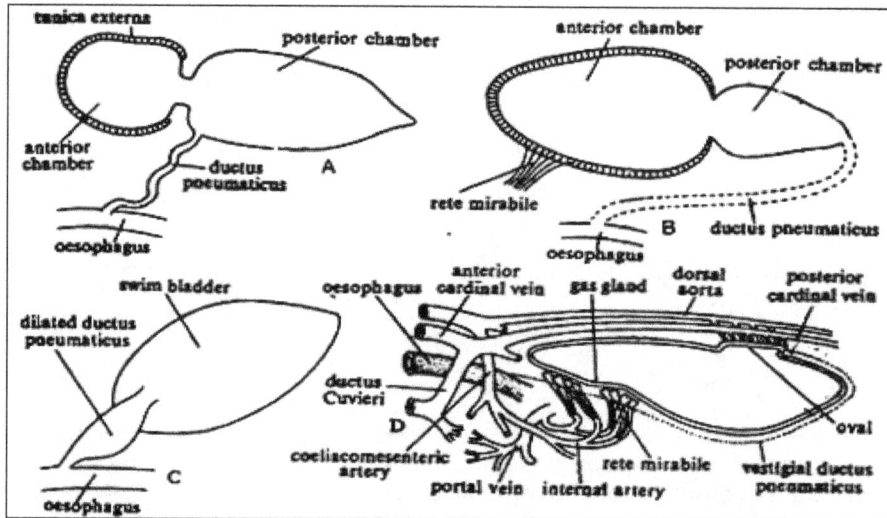

Variations in the structure of swim-bladder in fishes. A. Catostomus. B. A typical physocestous type of
swim-bladder with anterior and posterior chambers. C. Transitional swim-bladder in eel. Noes the dilatation
of the ductus pneumalicus. D. A physcefistous Innen-bladder with oval. Note the circulatory pathways.

The swim-bladder in Polypterus (bichir) represents the primitive condition. It is a bilobed sac with
two unequally developed lobes. The left lobe is shorter and the right lobe is longer. The bilobed sac
opens on the floor of the pharynx through a slit-like glottis. The glottis is provided with muscular
sphincter. The internal lining of the bladder is smooth and partly ciliated.

The lack of alveolar sacculations and the presence of muscular walls are the two noted feature in
the swim-bladder of Polypterus. The walls of the bladder are highly vascular and are lined by two
layers of striated muscle fibres.

The bladder is supplied by a pair of pulmonary arteries arising from the last pair of pulmonary
arteries arising from the last pair of epibranchial arteries and the corresponding veins enter into
the hepatic vein below the sinus venosus.

In the dipnoans, the swim-bladder is called the lung and the inner walls are produced into numerous alveoli. The swim-bladder resembles the tetrapod lungs both structurally as well as functionally. In Neoceratodus it is single- lobed, while in Protopterus and Lepidosiren it is bilobed.

Other details regarding the structural construction, blood' and nerve supplies have already been
dealt in the biology of the lung-fishes.

In Sturgeons (Acipenser), the swim-bladder is short and oval in shape. The ductus pneumaticus enters the bladder ventrally and it opens into the gut posterior to the pharynx. The glottis is
lacking and the opening into the oesophagus is closed by the simple constriction of the ductus
pneumaticus.

The walls of the bladder are fibrous and thick but the inner walls are smooth. In Acipenser,
both the left and right lobes develop from the dorsal side of the oesophagus in the embryonic
stage, but the left one becomes completely obliterated and right one gives rise to the adult
swim-bladder.

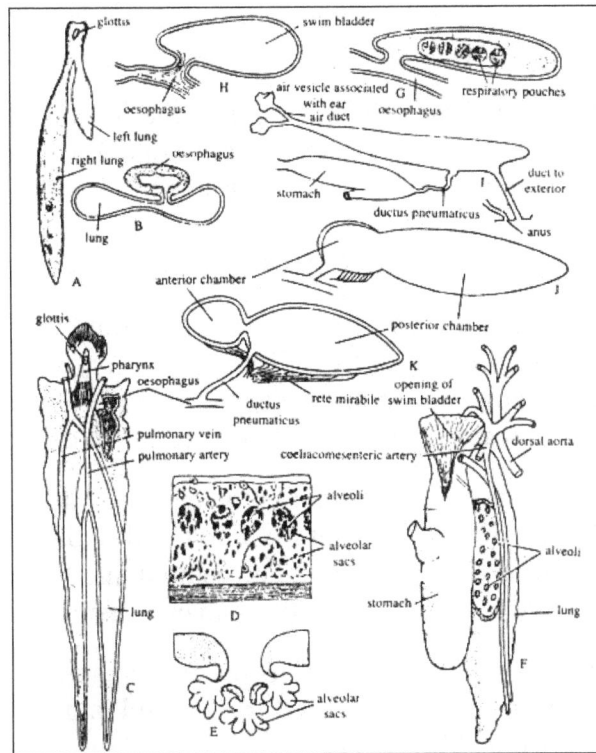

Figure shows variations in the structure of swim-bladder in fishes. A. Swim-bladder in Polypterus. The swim-bladder is modified as 'lung'. Note that the 'lung' is smaller in size. B. Sectional view of the swim-bladder and oesophagus in Polypterus showing their relationship. C. Swim-bladder of Polypterus. The swim-bladder is modified as the 'lung' D. A portion of the internal cavity of the 'lung' of Polypterus is enlarged to show the location of the alveoli. E. Showing a single alveolus of the 'lung' of Polypterus. F. Swim-bladder of Gymnarchus and its relative position. The swim-bladder is regarded as 'lung'. A portion of the 'lung' is removed to show the alveoli. G. Swim-bladder of Arnia and Lepisosteus. A portion is cut open to show the internal structures. H. Swim-bladder of Acipenser. I. Swim-bladder of Clupea harengus. J. Swim-bladder of Esox. K. Swim-bladder of a Cypnnoid fish.

In Amia and Lepisosteus, the swim bladder is an unpaired sac extending nearly the entire length of the body cavity. In both the cases rudiment of the left lobe appears during development but persists only for a short time. The ductus pneumatics opens into the oesophagus posterior to the pharynx through a dorsal slit-like glottis.

The walls are highly vascular and exhibit sacculations resembling the pulmonary alveoli. The sacculations or the respiratory pouches are arranged in two lateral rows. As regards the development of sacculations the swim-bladder of Lepisosteus is more advanced than that of Amia. There are some more minor differences regarding the supply of blood.

The swim-bladder in Amia gets arterial blood from the pulmonary arteries, while that of Lepisosteus gets arterial branches from the dorsal aorta. The blood from the bladder is returned by the left ductus Cuvieri in Amia and by the right post-cardinal in Lepisosteus.

Gymnarchus presents an intermediate stage where the efferent branchial arteries from the third and fourth gill-arches join to form a common root for the emergence of the pulmonary and coeliacomesenteric arteries. Amongst the dipnoans, the swim-bladder of Neocertatodus resembles that of Lepisosteus. The walls are sacculated and act as the lung'.

In Clupea harengus, the ductus pneumaticus opens into "the fundus of the stomach and there is a second duct from the posterior part of the swim-bladder opening to the exterior near the anus. Similar posterior opening is present in Pellona, Caranx, Sardinella.

Modifications of Physoclistous Condition

The swim-bladder in all teleosts begins as a physostomous type but in an adult condition the ductus pneumaticus gets degenerated to become a physoclistous type. A typical physoclistous swim-bladder consists of a closed sac having two compartments—an anterior and a posterior. These two compartments are intercommunicated through an aperture called ductus communicans.

The opening and closure of this aperture is regulated by circular and radiating muscles which act as the sphicter. The anterior chamber is formed by circular and radiating muscles which act as the sphincter. The anterior chamber is formed by the enlargement and forward growth of the budding swim-bladder, while the posterior chamber develops as an enlargement of the ductus pneumaticus.

This typical structural plan is modified in certain forms. The posterior chamber with retia Mirabella becomes flattened almost to the point of obliteration and is designated 'oval' as seen in the families like Myctophidae, Percidae, Mugilidae.

The oval is a thin-walled highly muscular area specialised for the reabsorption of gases. The opening of the oval is guarded by circular and longitudinal muscles. This device is of great significance for the fishes undergoing rapid vertical movements.

Histological Modifications

The morphological modifications of the swim-bladder are accompanied by histological modifications in different fishes, the swim-bladder acts as a hydrostatic organ. It helps fishes to sink or ascend to various depths by altering the gas content in the bladder. In fishes having open ductus pneumaticus, the volume of gas content in the bladder can be changed by swallowing or removing air from the bladder.

But in some physostomous and all physoclistous fishes this process of gas transference is done directly from the blood stream. Inside the bladder there is an oxygen-producing device and an oxygen-absorbing device. The swim bladder is a vascular structure but the degree of vascularization varies in different teleosts.

In some species of the families Clupeidae and Salmonidae the capillaries are uniformly present all over the swim-bladder, but in most cases these highly vascular interlacing and tightly packed capillaries form a mass called rete mirabilis. The anterior chamber of swim bladder shows the tendency to become differentiated into oxygen-producing area called red body.

The oxygen is produced by the reduction of the oxyhaemoglobin in the erythrocytes when brought into close contact with the secreting epithelial cells of the gas gland. The red body consists of internal oxygen-secreting cells (gas gland) and supplied by the blood vessels from the retia Mirabella (sing, rete mirabilis).

It forms a complicated structure where the arterial and venous capillaries communicate only after reaching the gas gland. The most primitive condition is observed in Pickerel where the gland is covered by thick glandular epithelium which is thrown into a number of folds. In eels and some other fishes, the red bodies are non- glandular in nature but serve the same physiological function.

The red gland is supplied with blood from the coeliac artery and is returned to the portal vein. The activity of the red gland is controlled by the vagus nerve. In the fishes with functional ductus pneumaticus the gas glands are absent but in eels this function is taken up by the red gland. In the physoclistous fishes, the anterior region is modified for gas production and the posterior region or chamber is specialised for the absorption of gas into the blood. The posterior chamber becomes excessively thin- walled to facilitate gas diffusion.

Beneath the walls, the gas is absorbed directly into the blood. The formation of the oval in some fishes is a special development for the absorption of gas. The wall of the oval is very thin and highly vascular. Through this epithelial lining oxygen can easily pass to the network of vessels. This gas absorbing region receives blood supply from the dorsal aorta and the blood is returned to the post cardinal vein. The activities are governed by the sympathetic nerves.

The histological differentiation for the gas production and gas absorption is a very significant achievement in fishes. The gas produced by the red body is mostly oxygen and this oxygen is readily absorbed or diffused from the swim-bladder directly into the capillaries. The oval is modified for gas absorption in many fishes.

By the alternate process of gas production and gas absorption, the internal pressure and volume of the gas content inside the swim-bladder can be increased or decreased. The red body is usually confined to the anterior chamber, but in fishes where the anterior chamber becomes secondarily associated with the auditory function, the gas gland may be confined to the posterior chamber.

Shape and Size of Swim-bladder

The swim-bladder varies extensively in shape and size. In Umbrina, it is oval shaped and without any appendage. In Atractoscion, it gives off only one pair of simple diverticula that extends from the anterior side. In Kathala, the swim-bladder develops a pair of appendage extending in front of transverse septum into head.

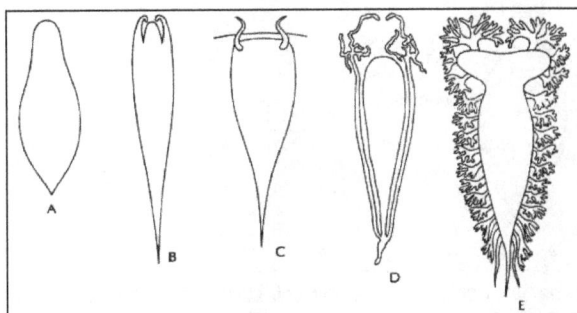

Variations in the shape of swim bladder. A. Umbrina.
B. Afractoscion, C. Kathala, D. Otolithoides. E. Johnius.

In some forms it gives off many branched diverticula. In many fishes, the anterior prolongations of the swim-bladder come into close contact with the wall of the space containing the internal ear. In

Clupea, the narrow anterior end of the swim-bladder enters into a canal in the basioccipital of the skull and divides into two slender branches.

The anterior end of each branch dilates to form a round swelling and lies in close contact with the internal ear. A more or less similar condition is observed in Tenualosa ilisha. In many fishes finger-like diverticula develop from the swim-bladder.

In Gadus, a pair of diverticula originating from the anterior part of the bladder project into the head region. In Otolithus, each anterolateral end of the swim-bladder gives rise to an outgrowth which sends one anterior and a posterior horn.

In Otolithoides, the appendages attached to posterior end of bladder and at least the main part lying parallel to the bladder. In Corvina lobata, many such branched diverticula develop from the lateral walls of the swim-bladder. In Johnius, it is hammer-shaped with 12 to 15 pairs arborescent appendages, the first branching in the head and the posterior tip are highly pointed.

Usually in most cases, the swim-bladder is divided transversely into an anterior and a posterior chamber as seen in cyprinoids, Esox , Catostomus, Pangassius, Corvina, etc. But the longitudinal division of the swim-bladder is rare.

In Arius the swim-bladder is splitted longitudinally. In Notopterus, a longitudinal septum divides the swim-bladder into two lateral chambers. Due to the presence of septum or septa, the internal cavity of the swim-bladder is either completely or partially divided.

Weberian Ossicles

The perilymphatic sac and the anterior end of the swim-bladder are connected by a series of four ossicles, which are articulated as a conducting chain.

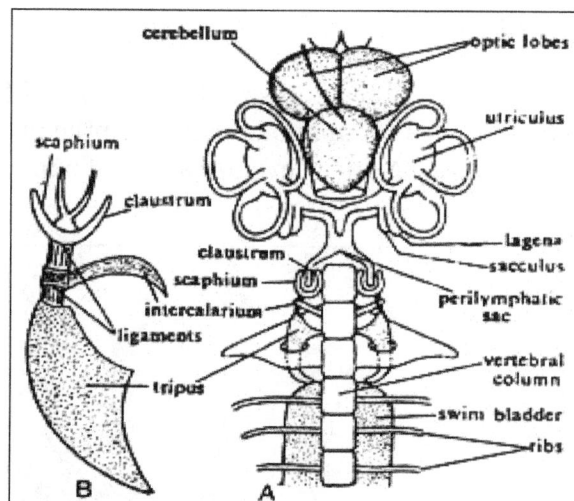

A. Weberian ossifies and their relation with other structures in
Catastomus. B. Showing the different parts of Weberian ossicles.

Of the four, the tripus, intercalarium and scaphium actually form the chain, while the fourth one, claustrum lies dorsal to the scaphium and lies in the wall of posterior prolongation of the perilymphatic sac. The function of these ossicles is controversial.

It is regarded that the Weberian ossicles either help to intensify sound vibrations and convey these waves to the internal ear of help to understand the state of tension of air pressure in the bladder and transit changes of such pressure to the perilymph to set up a reflex action. There are various views regarding the actual process of derivation of these ossicles.

De Beer and Watson regarded that these are detached or modified processes of the first three anterior vertebrae. As regards the actual mode of origin of the four ossicles there are differences of opinion.

The claustrum is regarded to be modified interspinous ossicle or modified spine of first vertebra or modified neural arch of first vertebra or modified intercalated cartilage or modified neural process of first cartilage.

The scaphium is considered to be the modified neural arch of the first vertebra or modified rib of the first vertebra or derived from the neural arch of the first vertebra and also from the mesenchyme.

The intercalarium is derived from the neural arch and transverse process of the second vertebra or from the neural arch of the second vertebra and also from the ossified ligament or from the neural arch of the second vertebra only.

The tripus is formed from the rib of the third vertebra and the ossified ligament or from the transverse process of the third vertebra along with ossified wall of the swim-bladder or from the transverse process of the third vertebra and the ribs of third and fourth vertebrae.

Functions of Swim-bladder

The swim-bladder in fishes performs a variety of functions:

Hydrostatic Organ

It is primarily a hydrostatic organ and helps to keep the weight of the body equal to the volume of the water, the fish displaces. It also serves to equilibrate the body in relation to the surrounding medium by increasing or decreasing the volume of gas content.

In the physostomous fishes the expulsion of the gas from the swim-bladder is caused by way of the ductus pneumaticus, but in the physoclistous fishes where the ductus pneumaticus is absent the superfluous gas is removed by diffusion.

Swim-Bladder acts as Adjustable Float

The swim-bladder also acts as an adjustable float to enable the fishes to swim at any depth with the least effort. When a fish likes to sink, the specific gravity of the body is increased. When it ascends the swim-bladder is distended and the specific gravity is diminished. By such adjustment, a fish can maintain equilibrium at any level.

Swim-bladder Maintains Proper Centre of Gravity

The swim bladder helps to maintain the proper centre of gravity by shifting the contained gas from one part of it to the other and this facilitates in exhibiting a variety of movement.

Swim-bladder helps in Respiration

The respiratory function of the swim-bladder is quite significant. In many fishes living in water in which oxygen content is considerably low, the oxygen produced in the bladder may serve as a source of oxygen. In a few fishes, especially in the dipnoans, the swim bladder becomes modified into the 'lung'. The 'lung' is capable of taking atmospheric air.

Swim-bladder as Resonator

The swim-bladder is regarded to act as a resonator. It intensifies the vibrations of sound and transmits these to the ear through the Weberian ossicles.

Production of Sound

The swim-bladder helps in the production of sound. Many fishes, Doras, Platystoma, Malapterurus, Trigla can produce grunting or hissing or drumming sound. The circulation of the contained air inside the swim-bladder causes the vibration of the incomplete septa.

The sound is produced as the consequence of vibration of the incomplete septa present on the inner wall of the swim-bladder. The vibrations are caused by the movement of the contained air of the swim-bladder.

Sound may also be produced by the compression of the extrinsic and intrinsic musculature of the swim-bladder. Polypterus, Protopterus and Lepidosiren can produce sound by compression and forceful expulsion of the contained gas in the swimbladder. In Cynoscion male, the musculus sonorificus probably helps in compression.

Kidney

Throughout all organisms, the function of the kidneys generally is to ensure that osmoregulation is kept constant. That is that the water and salt balances within the bodies of the organisms are kept at the appropriate level needed for the organism to survive. When it comes to fish especially, the kidneys are designed to interact with the internal body fluids on how they interact with external water fluids. Therefore the kidneys function slightly differently between those fish that live in saltwater and those that live in freshwater.

The kidneys in fish are slightly different to those found in mammals. They are smaller, stretched and do not resemble beans as such. Generally, within fish, the kidney is located towards the very middle of the fish both lengthwise and height wise.

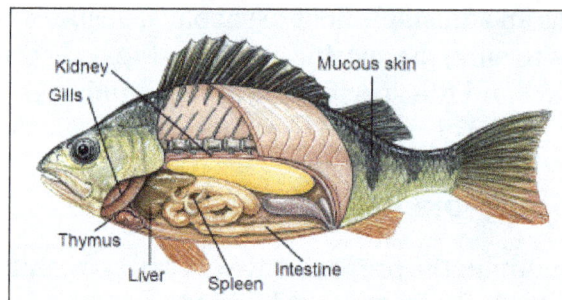

Freshwater Fish

Within freshwater fish, the kidney is responsible for preventing excess solute loss as they contain more salts within their blood than the water located around thier bodies. Due to this concentration gradient, water will naturally diffuse into the fish through osmosis causing large amounts of water to build up inside. Therefore, the kidneys in this type of fish increase the amount of water that passes out in their urine and actively reabsorb the salts that would pass out as well to maintain that balance. This therefore results in the production and excretion of large amounts of dilute urine. This urine is so dilute that it is almost completely composed of water.

Saltwater Fish

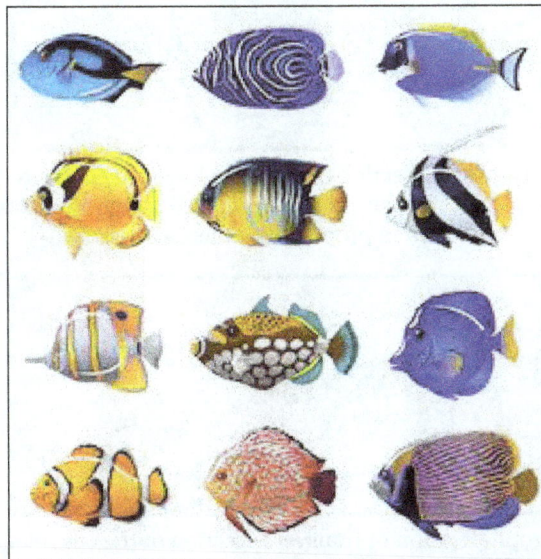

Saltwater fish are almost the complete opposite when it comes to kidney function as the water and liquid around them is highly concentrated with salts and other solutes. This means that the water within the fish is constantly being 'sucked out' of them into the surrounding water through osmosis. This is because the fluids inside their bodies are less concentrated than the saltwater around their bodies. Therefore the fish must counteract this process through drinking water and the use of their kidneys. By constantly in taking water, the benefits include that the fish is constantly replacing the water that is sucked out of it however, this can lead to an extreme intake of salts and build-up of harmful substances within the fish. These salts are removed through the use of the kidneys whereby they help the fish retain water and actively excrete salts to produce very concentrated urine. This also means that saltwater fish do not urinate as much excrement as does freshwater fish.

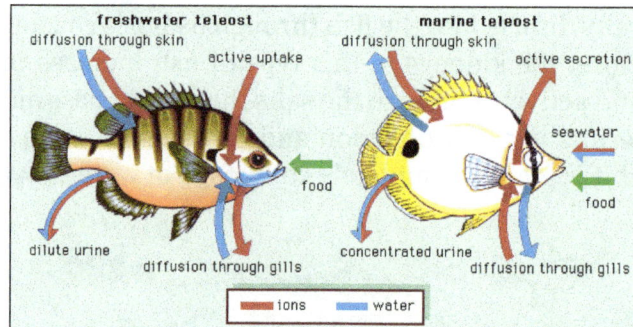

Stomach

The gut is made up of usual four layers, viz. serosa, muscularis externa, sub-mucosa and mucosa. The serosa is made up of loose connective tissue. Next to serosa is muscularis externa. It is distinguished into an outer longitudinally arranged muscle fibres, whereas the inner layer is composed of circular muscle fibres.

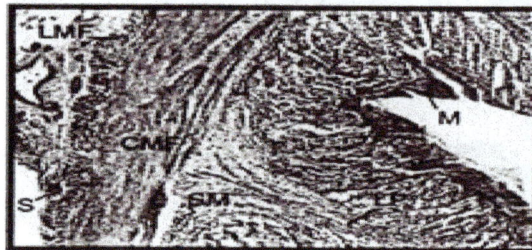

Microphotograph of transverse section of the oesophagus of clarias to batrachus showing low layers X 80. CMF. circular muscle fibre; LMF, longitudinal muscle. fibre; L P. lamina propria; M. mucosa. S,serosa.

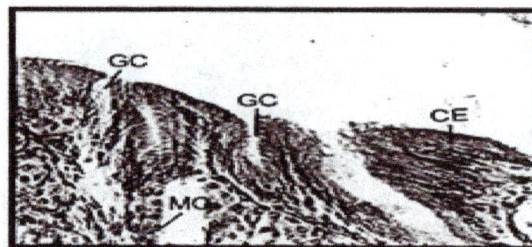

In figure, microphotograph of transverse section of the oesoplagus of Clarias batrachus showing mucosal folas X320. CE, columnar eplthelium; CC goblet cell; MC, mucus cell.

The submucosa is made up of loose connective tissue, blood vessels and capillaries. The submucosa is followed by innermost mucosa, which is divisible into lamina propria and epithelial layer. The lamina propria is vascular and is made up of areolar connective tissue.

The epithelial layers which lines the lumen of the gut is made up of columnar epithelium and is thrown into deep mucosal folds. Mucosa comprises of various glands. The stomach possesses broad mucosal folds subdivided into primary and secondary folds. The mucosa contains gastric glands.

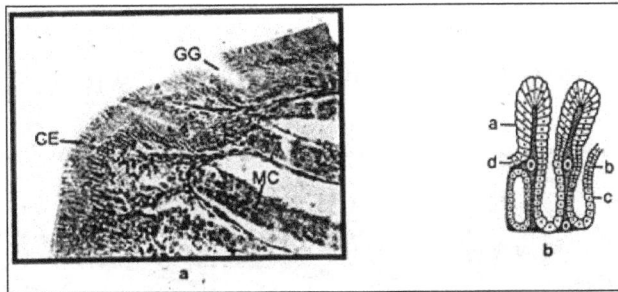

Mucosal, folds of stomach. (a) Microphotograph of transverse section of stomach of Clarias, batrachus X 320. (b) Diagram of mucosal epithelium of the stomach. (c) gastric epithelium; columnar epithelium, (d) glandular CE, columnar epithelium; GC gastric gland, MC, mucus cells.

The submucosa is reduced having bundles of longitudinal muscle. Circular muscle fibre coat is well developed. Serosa is thin.

The submucosa is well developed followed by thick coat of circular muscles, which is surrounded externally by longitudinal muscle fibres. The serosa thin comprises flattened epithelial cells. In the intestine, the mucosal, folds are produced into prominent slender folds called villi, which have intestinal glands.

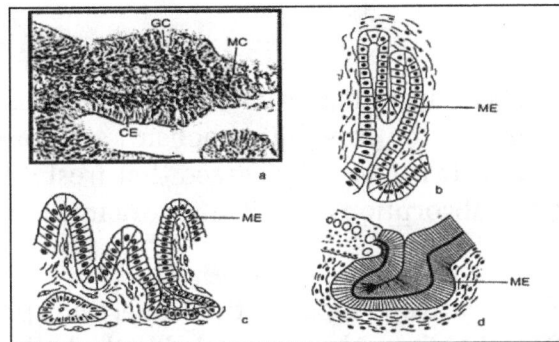

Diagrams of transverse section of intestine. (a) microphotograph of duodenum of clarias batrachus. X 320. (b), (c), (d), diagram of mucosal folds of intestine. C.E, columnar epithelium; GC. Goblet cell; MC, mucus cell; ME, mucosal epithelium.

The submucosa extends into villi forming lamina propria. The circular and longitudinal muscle layers are comparatively thin than in the stomach.

The rectum has short and flat mucosal folds, provided with large number of mucous cells than the intestine. The muscular coat is thick.

Microphotograph of rectum of clarias batrachus showing four layers and mucus cells. X320. CE, columnar epithelium; CMF, circular muscle fibres; LMF, longitudinal muscle fibres, MC, mucus cells; S, serosa.

Innervation of Gut

The alimentary canal of fishes is innervated by sympathetic and parasympathetic components of autonomic nervous system.

Microphotograph of transverse section of oesophagus of clarias batrachus showing nerve, fibres in the submucosa and lamina propria. X 320, NE, nerve Fibre; LP, lamina propria, SM, Submucosa.

Presence of nerve plexus in various part of the alimentary canal has been reported by Tembhre and Kumar and Nicol. The presence of neurotransmitter acetylcholine in intestinal bulb and intestine of fish has been reported both histo- chemically and biochemically.

Metabolism

The proteins, carbohydrates, fats, most minerals and vitamins are essential dietary requirement of fishes. They must be taken in diet for growth (anabolism) and for energy (catabolism). They take mineral from ambient water. It is generally agreed that freshwater as compared with marine fishes have relatively higher absorption capacity of inorganic ions because of the surrounding water.

Protein is needed in the diet for growth and repair of the tissue. Body protein consists of long chain amino acids. Only twenty different amino acids are needed in the body for the synthesis of protein molecule. Out of these twenty amino acids in human beings, 8 are essential amino acids.

They must be present in the diet, body cannot synthesize them. In fishes, 10 amino acids are essential. Arginine and histidine are the two amino acids which are extra and the rest 8 are similar to that of human being.

Table: The amino acids are as follows.

Non-essential amino acid	Essential acids
Alanine	Arginine
Aspartic acid	Isoleucine
Cysteine	Leucine
Glutamic acid	Lysine
Glutamine	Methionine
Glycine	Phenylalanine
Proline	Threonine

Serine	Tryptophane
Tyrosine	Valine

Digestion of Food

For the digestion of protein, the following enzymes are required in vertebrate series.

1. Pepsin (Stomach of carnivorous fishes),

2. Trypsin (Intestine (alkaline medium), pancreas, intestinal caecae),

3. Chymotrypsin,

4. Erypsin (Collection of peptidases are known as erypsin, found in intestine).

Digestion of Proteins

The fishes which possess stomach are generally carnivorous and secrete pepsin enzyme from gastric mucosa. The pepsin is a protease enzyme, i.e., it can break down protein. The optimal activity is carried out at a pH 2 to 4, so HCl is required for making low pH. HCl is secreted by the gastric mucosa in carnivorous fishes creating the low pH.

Both cholinergic and adrenergic nerves are present in the stomachs which stimulate the secretion of gastric juices. The secretion of gastric juices (acid secretion and pepsin) depends upon temperature. At 10°C the gastric secretion increases to three to four folds.

The trypsin enzyme is present in the extract of pancreas of some Elasmobranchs such as Mustelus cartarias, Littoralis and Squalus. The trypsin is secreted by exocrine pancreatic tissue which may be concentrated in a compact organ as in mackerel (Scomber) or diffusely located in mesentric membranes surrounding the intestine and liver. It is also secreted by hepatopancreas.

The inactive form of this enzyme trypsinogen is known as zymogen. It is to be converted into active enzyme, i.e., trypsin by an enzyme enterokinase. The enterokinase enzyme is exclusively secreted by intestine of fish.

In the Cyprinids, stomach-less fish, protease compensation is supplemented by some intestinal enzyme known collectively as erypsin. Pepsin is absent in stomach-less fishes because of the absence of true stomach.

The intestine secretes amino-peptidases. These act on terminal amino acid called as exopeptidases and those act on central bonds are called as endo-peptidases. Vitamins are essential constituent of the diet and a large number of vitamin deficient syndrome are noticed in fish.

Vitamin Deficiency Syndrome in Fishes

- Vitamin: Symptoms in Salmon, Trout, Carp, Catfish.

- Thiamine: Poor appetite, muscle atrophy, convulsions, instability and loss of equilibrium, edema, poor growth.

- Riboflavin: Corneal vascularization, cloudy lens, haemorrhagic eyes, photophobia, dim vision, incoordination, abnormal pigmentation of iris, striated constrictions of abdominal wall, dark coloration, poor appetite, anaemia, poor growth.

- Pyridoxine Acid: Nervous disorders, epileptiform fits, hyper irritability, ataxia, anemia, loss of appetite, edema of peritoneal cavity, colorless serous fluid, rapid postmortem rigormortis, rapid and gasping breathing, flexing of opercles.

- Pantothenic: Clubbed gills, prostration, loss of appetite, necrosis and scarring cellular atrophy, gill exudate, sluggishness, and poor growth.

- Inositol: Poor growth, distended stomach, increased gastric emptying time, skin lesions.

- Biotin: Loss of appetite, lesions in colon, coloration muscle atrophy, spastic convulsions, fragmentation of erythrocytes, skin lesions, poor growth.

- Folic Acid: Poor growth, lethargy, fragility of caudal fin, dark coloration, macrocytic anemia.

- Choline: Poor growth, poor food conversion, haemorrhagic kidney and intestine.

- Nicotinic Acid: Loss of appetite, lesions in colon, jerky or difficult motion, weakness, edema of stomach and colon, muscle spasms while resting, poor growth.

- Vitamin B_{12}: Poor appetite, low haemoglobin, fragmentation of erythrocytes, macrocytic anaemia.

- Ascorbic Acid: Scoliosis, lordosis, impaired collagen formation, altered cartilage, eye lesions, haemorrhagic skin, liver, kidney, intestine, and muscle.

Digestion of Carbohydrates

The term carbohydrates was originally derived from the fact that large bulk of compounds being described fit in the empirical formula Cn (H_2O)n. Though formaldehyde, acetic acid and lactic acid fulfill the formula requirement, they are not carbohydrates.

The useful definition of carbohydrate might be poly-hydroxy-aldehyde and ketones and their derivatives. This would include de-oxy-sugars, amino sugars and even sugar alcohols and acids. The enzymes which break down the carbohydrates in the gut of fishes are carbohydrates.

They are as follows:

- Amylase
- Lactase
- Saccharsases/sucrase
- Cellulase

The most important enzyme is amylase which acts on starch (amylum) and which breaks down to maltose and then to glucose by the process of digestion. In human beings, the amylase is secreted from salivary glands and pancreas.

The amylase is secreted from the pancreas in carnivorous fishes but in herbivorous fishes, the presence of this enzyme is reported from the whole gastrointestinal tract as well as from pancreas. Researches on the carbohydrates of fishes have largely been confined to the identification of amyloclastic activity.

The pancreatic extract of Raja, an Elasmobranch, Scyllium have clearly shown the amylase activity in the pancreatic juice. Tilapia (Sarotherodon mossambicus), which is herbivorous, the amylase is present throughout the alimentary tract. In Rasbora daniconius, Saxena; Kothari reported amylase in the intestinal bulb, duodenum and ileum.

In reviewing the literature, it is clear that pancreas (hepatopancreas) is the chief site for the production of amylase, although the intestinal mucosa and intestinal caecae represent additional production site in various species. The enzymatic activity of these caecae is known to be lower as compared to the intestine under normal condition.

$$\text{Starch} \xrightarrow{\text{Amylase}} \text{Maltose} \xrightarrow{\text{Maltase}} \text{Glucose}$$

In those fishes in which sucrose has been reported the effect is as follows:

$$\text{Sucrose} \xrightarrow{\text{Sucrase}} \text{Glucose} + \text{Fructose}$$

How galactose is further hydrolyzed is not clear in fishes. Blood glucose is converted with the aid of insulin, to muscle glycogen. Although clear details are wanting, but excess of glucose enters the blood from digestive tract, the surplus is converted to glycogen in liver.

Endocommensal Bacteria

Lagler stated that in fishes such as menhaden (Brevoortia), silverside (Menidia) and silverperch (Bairdiella) possess endo-commensal bacteria containing an enzyme, the cellulase, which breaks down the cellulose plant material.

Cellulose of plant material containing starch could be broken down to glucose by cellulase enzyme of these bacteria instead of passing out through faeces.

Fat Digestion

The lipids are organic substances insoluble in water but soluble in organic solvents like chloroform, ether and benzene. They form important dietary constituents on account of high caloric value and the fat soluble vitamins and the essential fatty acids contained in them.

The main enzyme which acts on this lipid is lipase. The pancreas is also the primary site of lipase production. Vonk found lipase in the pancreas of trout but also found this enzyme in mucosa of fishes.

Gastrointestinal Hormone

The mucosa of gastrointestinal tract of man possesses four hormones. They are secretin, cholecystokinin (CCK), gastrin and gastric inhibitory peptide. In each case the hormone is released into the blood stream by the gastrointestinal endocrine cells and as it circulates throughout the body, it is bound by receptors on the plasma membrane of the target cells.

In the teleost, the presence of gastrin and cholecystokinin are reported and are secreted by intestinal endocrine cells which are dispersed and are not grouped in clusters. The CCK affects the oxyntic cells and inhibits further gastric secretion in bony fishes.

Somatostatin is present in the stomach and pancreas of fishes. They are called as paracrine substances. It differs from hormone because it diffuses locally to the target cells instead of releasing into the blood. This inhibits other gastrointestinal and pancreatic islet endocrine cells.

The occurrence of VIP (vesoactive intestinal peptides) and PP (pancreatic peptide) have been reported in S. aurotus and B. conchonius in the gastric tract. These are classified as candidate hormone. These are gastrointestinal peptides whose definite classification as hormone or paracrine has not been established.

These are designated as candidate or putative hormone. Pancreas secretes two important hormone, i.e., insulin and glucagon, insulin is secreted from β-cells while glucagon is secreted by α-cells.

In addition to acetylcholine (not peptide) known to be in nerve fibres of gastrointestinal tract present gastric VIP (Vasoactive Intestinal Peptides) and somatostatin, met-enkephelin and substance P are reported in the teleost fishes.

Absorption

Inorganic ions uptake in various regions of the alimentary canal in fishes and their subsequent distribution and localization have been reported. The iron (Fe $^{++}$) ions are absorbed through intestinal columnar cells and then pass into the portal blood as Fe^{++} binding protein transferitin.

The calcium is absorbed by the intestinal submucosal blood vessels. Probably. Ca^{++} after entering blood vessels in the intestinal region reaches finally in the hepatocytes where it is stored in association with vitamin D depending upon Ca^{++} binding protein.

Regarding dietary calcium and phosphorus absorption, Nakamura and Yamada, Nakamura and Sinha and Chakraborti reported calcium and phosphorus in the digestive tract of Cyprinus carpio and Labeo rohita. In teleosts, ambient water also serves as external source of various dissolved minerals besides food.

Heart

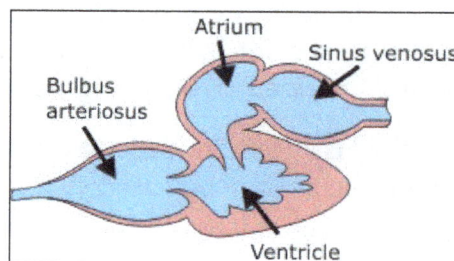

The heart is the pump that generates the driving pressure for the circulation of blood. The fish heart has one atrium and one ventricle; this is in contrast to the human (mammalian) heart that has two separate atria and two separate ventricles. In the fish heart, two other chambers can also be found: the sinus venosus and the bulbus arteriosus.

Structure of Heart

The heart of fishes is known as branchial heart, because its main function is to pump venous blood to ventral aorta into gills (branchial) and then to somatic vasculature. Thus branchial and systemic vascular beds are arranged in series with heart.

Apart from heart, heart-like organs are present only in Agnatha (Myxine and Petromyzon). The heart of fishes consists of four chambers, a sinus venosus, an atrium, a ventricle and a conus or a bulbus arteriosus.

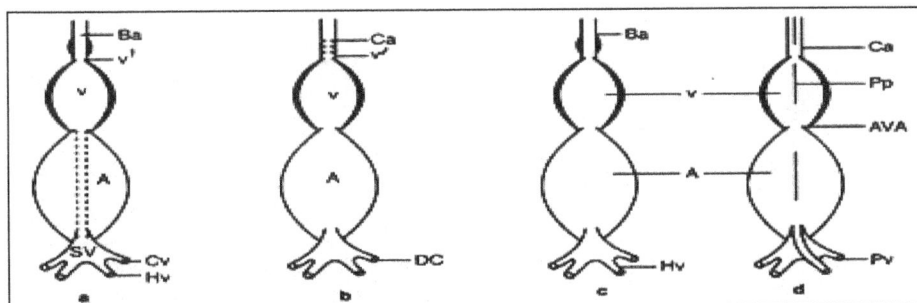

Diagrammatic sketch of structure of the heart fishes.

In figure, (a) lamprey heart showing bulbus arteriosus. (b) Cartilaginous fish heart Showing Conus arteriosus having many valves. (c) heart of bony fish showing conus and buldus arteriosus. (d) dipnoan heart showing partial septum in the atira and ventricle. AVA, atrioventricular aperture; A, atrium; Ba, bulbus arteriosus; Ca, conus arteriosus; Cv, cuvieri vein; DC, ductus cuvieri; Hv, hepatic vein; Pv, post caval; SV sinus venosus v, valve. V', Ventricle.

In elasmobranchs the fourth chamber is designated as conus arteriosus whereas it is known as bulbus arteriosus in teleost, a specialized ventral aorta in teleosts.

The distinction between the two is that the conus consists of cardiac musculature similar to the ventricle and is generally provided by a large number of valves arranged in successive rows while bulbus arteriosus comprises only smooth muscle fibres and elastic tissues.

The heart of Cyprinus carpio a teleostean fish, contains both conus and bulbus arteriosus. However, later workers held that in teleosts only bulbus arteriosus is present. Elasmobranch and aganthan have conus arteriosus instead of bulbus arteriosus.

Heart Rate and Stroke Volume

The heart performance basically depends on two factors; the heart rate and the stroke volume. At each heartbeat, the ventricle pumps out blood. The volume is termed stroke volume and the time of the heart beat is known as heart rate.

These are controlled both by aneural factors such as extent of cardiac filling (Starlings law of the heart) or circulatory substances (hormones) and by the innervation of the cardiac pacemaker and muscle.

The fish atrium is filled by suction created by the rigidity of the pericardium and surrounding tissue. Venous blood return to the atrium is aided by ventricular contraction in systole which causes a fall in intra-pericardial pressure that is transmitted through the thin wall of the atrium to create an aspiratory or via a fonte effect.

It is contrary to the situation in mammals where the central venous pressure determines the atrial filling during diastole (vis a tergo, driving force from behind.)

Sinus Venosus

The sinus venosus is not an active part of the heart although pacemaker properly starts in this chamber.

Distribution of pacemaker in the sinus venosus of eel. The Pacemaker area is shown by strips. (a) suggested by bielig. (b) As suggested by Grodzinski. A, atrium; AVp atrioventricular pad; Ba bulbus arteriosus; Ca, conus arteriosus; Dc, ductus Cuvieri; Sv, sinus venosus; v, ventricle.

It is actually a continuation of the venosus vessels and its main function is to receive blood and to pass it on to the atrium. Sinus venosus receives blood through two ducts Cuvieri, hepatic veins pour the blood from liver. The ventral ductus Cuvieri receives blood from anterior and posterior cardinal veins.

The sinus venosus is distinguished histologically into tunica intima, tunica media and tunica adventitia. Normally the sinus venosus is purely amuscular in some fishes. The matrix of this chamber is made up of elastic and collagen fibres.

The muscles are restricted around the sinuatrial opening in circular fashion forming sinuatrial ring. The sinus venosus opens into atrium by a sinuatrial ostium, which is provided by two sinuatrial valves. Farrel and Jones reported single atrioventricular valve in teleost fishes.

Atrium

The atrium is a large muscular contractile chamber. It is situated dorsal to the ventricle in almost all fishes. In fishes, the atrium is also known as auricle, but actually the appendages of the atria are called as auricles. The atrium is undivided single chamber in elasmobranch and teleosts but in dipnoi, the atrium is partially divided by an incomplete interatrial septum.

Diagram of heart showing various histological layers of heart chabmbers. (a) Trout (teleost). (b) shark (elasmobranch). A atrium; Ap, apex; CA, conus arteriosus; Epi, epicardium; End, endocardium; Myo, Myocardium; PcM pacemaker region; Sv, sinus venosus; v,ventricle.

Pulmonary blood drains directly into the left side of the atrium, whereas the systemic venous blood is collected in the sinus venosus through ducti Cuvieri. The blood from the sinus venosus goes to the right side of the atrium.

Internally, the atrium is divisible into two parts, a sinuatrial canal and atrium proper. The former is rather thick-walled semi-cylindrical rigid tube and the latter is a thin-walled distensible spongy cavity. The significance and functional importance of this funnel is due to the pressure of blood in the sinus venosus and atrial filling.

The spongy portion of the atrium contains pectinate muscles. The trabeculae at the atrioventricular ostium form mesh-like network. When they contract, they pull the roof and sides of the atrium towards atrioventricular ostium. The atrial mass constitutes 0.25% of ventricular mass and 0.01-0.03% of body weight.

The atrium histologically is distinguished into epicardium, endocardium and myocardium. The endocardium is the innermost layer, lining the lumen of the atrium. The endothelial cells are flat with spheroid or more often elongated nuclei.

Atrioventricular Funnel

The atrium communicates with the ventricle through a tubular structure referred to as canalis auricularis or atrioventricular funnel. The atrioventricular opening is round and guarded by atrioventricular valves.

Regarding the disposition and number of AV valves in the heart of fishes in general and teleosts in particular are still much disputed. Generally, in teleosts two atrioventricular valves are present but Farrel and Jones described a single atrioventricular valve.

The atrioventricular valves in all the three genera of dipnoans, lungfish, i.e., Protopterus (Africa), Lepidosiren (South America) and Neoceratodus (Australia) are replaced by another structure known as atrioventricular plug.

The atrioventricular plug which guards the horseshoe-shaped atrioventricular opening, the functions is similar to atrioventricular valve. It is in the form of inverted cone with its apex pointing into the atrial lumen. It is projected dorsally with the atrial lumen and reaches up to pulmonalis fold and due to this; there is partial septation of the atrium.

It is made up of hyaline cartilage encircled by fibrous connective tissue. In Neoceratodus, the hyaline cartilage is absent and the plug is made up of fibrous connective tissue.

Ventricle

The teleost ventricle is tubular, pyramidal or sac-like in appearance.

It is relatively large muscular chamber. It is undivided in elasmobranch and teleost, but it is partially divided into left and right chambers by a muscular septum in Dipnoi. The muscular septum is posterior to the atrioventricular plug in all the three genera but extends anteriorly along the ventral surface in Lepidosiren. Its anterior and dorsal margins are free.

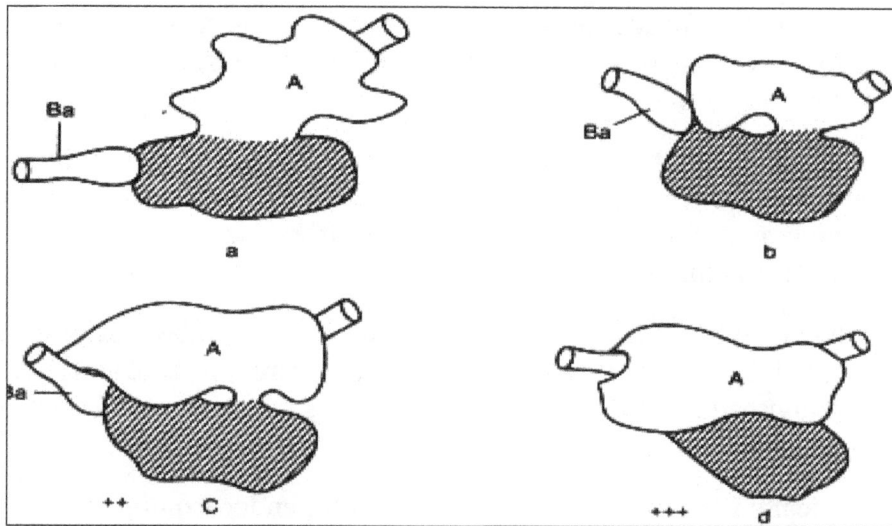

The degree of position of atrium to the bulbus arteriosus varies amongst teleost from a condition (-) where the two chambers are not in contact to a condition (+++) where the most anterior part of the atrium almost completely covers the bulbus arteriosus A, atrium; Ba, bulbus arteriosus.

Coronary Circulation

The working myocardium of the fish's heart, like other tissues require a blood supply to provide oxygen. There are two routes for oxygen supply and they are utilized to different degrees among fishes. Since the heart pumps venous blood, oxygen is available from the relatively oxygen poor venous blood that bathe the endocardial lining of the chamber.

In addition, an arterial supply of oxygen rich blood may be provided by the coronary circulation to the myocardium. All elasmobranchs and most active teleosts use both the venous and coronary oxygen supply to varying degrees.

Development of coronary circulation is generally associated with the relatively larger ventricle. In rainbow trout, Onchorhynchus mykiss, acetylcholine helps in the contraction of coronary arteries and there is predominantly relaxation with isoproterenol, epinephrine, nor-epinephrine and serotonin.

Coronary vascular resistance increases exponentially as coronary flow rate decreases. Coronary resistance was also influenced by cardiac metabolism and acclimatization.

Contractile Proteins

The available evidence suggests that properties of the contractile proteins from lower vertebrate are broadly similar to those found in the skeletal and cardiac muscle of mammalian species. However, adult cardiac muscles contain isotypes of myosin, tropomyosin and troponin which have distinct chemical structures and somewhat different properties from those found in skeletal muscle.

The complex orientation of fibres and the presence of a large proportion of non-muscle cells in cardiac tissues make it difficult to obtain multicellular preparations for the study of their contractile properties. The myosin isolated from fish and amphibians skeletal muscles are of unstable type which readily lose their ATPase activity of storage.

The fish actomyosin preparations are orders of magnitude more stable than corresponding myosin preparations. It is now of common belief that in common with myosin there have been selective modifications in the sequence of tropomyosin and troponin to permit efficient regulation of contraction at different body temperatures.

Pathology of Heart

The heart muscles are infected with bacteria and viruses. The bacterial infection is due to aero-monas and vibrios. They form colonies in the myocardium resulting that the endocardium becomes swollen and their nuclei become pycnotic. The viral infections commonly affecting the heart muscle are rhabdo-virus.

The infection causes myocardial necrosis resulting in inflammation in all the three layers, i.e., epicardium, endocardium and myocardium. The inflammation of cardiac muscle is known as myocarditis. A few reports deal with atrioventricular valve diseases. Like the higher vertebrates, the regeneration ability of cardiac muscle is nil and any injury or myocardial infarction develops into fibrous connective tissue.

The Cardiac Conducting System (Specialized Tissues)

The cardiac conducting system of homoiothermal vertebrates is responsible for the initiation and conduction of electrical impulse at right place and at right time. This system is also often called "Purkinje system" or "Specialized tissues".

In higher vertebrates this system is well developed and consists of a sinuatrial node (pacemaker muscle) situated in the right atrium, an atrioventricular node placed at the caudal end of the interatrial septum near the coronary sinus and the atrioventricular bundle placed above the inter-ventricular septum (bundle of His) and its two branches along with Purkinje fibres situated sub-endo-cardially both in the atria and the ventricle.

It has been unanimously accepted that the Purkinje fibres similar to those of higher vertebrates are absent in the heart of fishes. Whether the heart beat in fishes is generated by means of muscles or by nerves has not been clearly understood so far. The physiological investigations are few and also as controversial as morphological ones.

The cardiac beat is originated at the ostial part of the sinus and there are three groups of pacemaker in eel, while four groups were reported by Grodzinski. A few investigators found histologically specialized structures such as sinuatrial and atrioventricular plugs in the heart of fishes.

Presence of histologically specialized muscles which take lesser stain than the working cardiac muscles in fishes, have been reported in a few species. On the other hand, majority of workers denied the presence of histologically specialized tissues in any part of the heart of fishes.

Nodal Tissue

Keith and Flack and Keith and Mackenzie found nodal tissue at the base of venous valve. The criterion allowing a distinction of the nodal cells from other cardiac muscle cells in higher vertebrates is the relative poorness in myofibrils within cytoplasm as revealed by electron microscope.

This characteristic feature is reported in some part of the sinuatrial myocardium of loaches in catfish and trout.

There is no unanimity regarding the occurrence of nodal tissue in true histological sense, but almost all investigators in this field found heavy nerves and intimate nerve connection at the sinuatrial junction where pacemaker potential has been described.

There is a muscular continuity in various chambers of the heart and the chambers are not interrupted by nodes, bundles and Punkinje fibres.

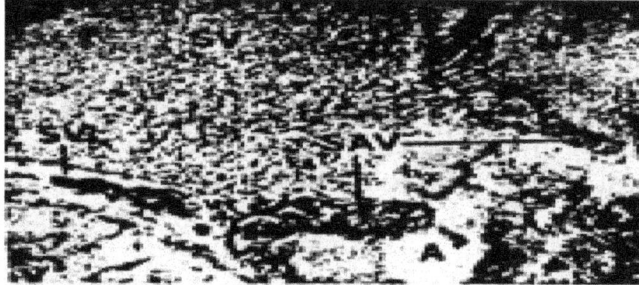

Photomicrograph of the heart of fish showing rich innervation at the sinuatrial junction. A, atrium; N, nerve; SV, sinus venosus; SAV, sinuatrial valve; SAL, sinuatrial Ligament; V, ventricle.

The distribution of nervous connection corresponds fairly accurately with electro-physiologically defined pacemaker region and it is, therefore, likely that there is a cholinergic vagal influence on pacemaker activity of dipnoans sp.

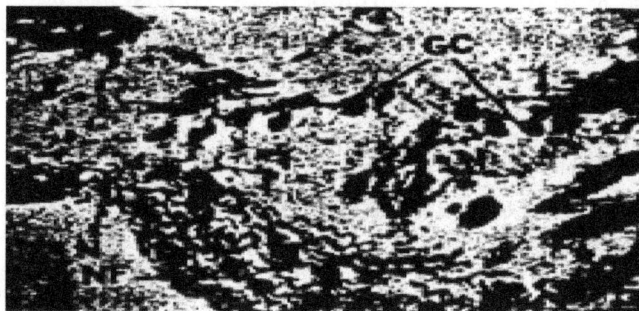

Microphotograph showing the presence of ganglion cells and ganglionated nerves in the heart of fish. GC, ganglion cell; N, nerve fibre.

Like other fishes dipnoans heart is also not provided with sympathetic innervation. From the sinuatrial region the contraction wave successively invades the atrium, the atrioventricular funnel and then to the ventricular myocardium.

It is generally believed that the cardiac conducting system of the heart of fishes is neither purely myogenic nor entirely neurogenic, but is the complex combination of the two.

Innervation of Heart

The heart of fishes is innervated by a pair of cardiac branch of vagosympathetic trunk, except in myxinoid heart, which receives no extrinsic innervation. Like other vertebrates, the heart is under autonomic control.

The autonomic nervous system in teleost is sympathetic and parasympathetic. There are no direct sympathetic nerves going to the heart. The vagus at its origin is parasympathetic (cranial outflow) but it receives postganglionic autonomic fibres from sympathetic chain into the head region.

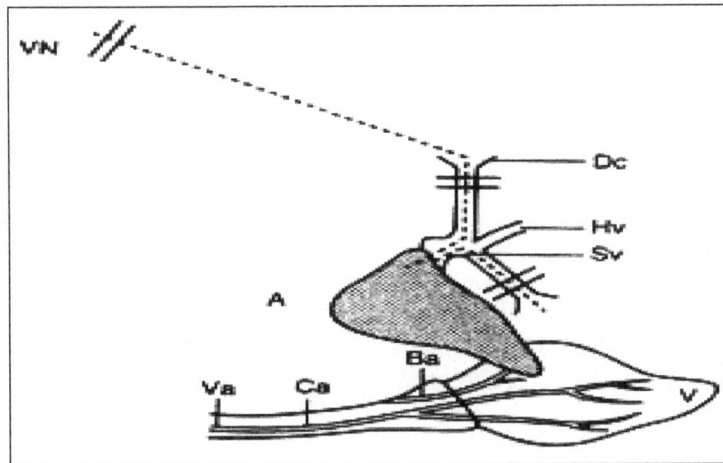

Diagram showing vagus nerve supplying to the heart through ductus cuvieri and also showing coronary artery supplying to the trout ventricle. A, arrium; Ba bulbus artereosus; Ca coronary artery, DC, ductus cuvieri; Hv, hepatic vein; Sv, sinus venosus; V; ventricle; VA ventral aorta; VN, vagus nerve.

The various chambers of the heart are richly innervated by both cholinergic and adrenergic nerve fibres. The different nerve endings (intra-cardiac mechano receptors) present in the heart when receive adequate stimulus transmit impulses to CNS.

This information then processed in CNS and then subsequently transmits impulses through autonomic (efferent) fibres to the heart which helps in atrial filling a cardio-ventilatory coupling. The heart of the fish, as of higher vertebrates, is under inhibitory control by cholinergic vagal fibres. Cholinergic nerves present in the heart secrete ACh, a neurotransmitter at their termination is essential for impulse transmission and action potential.

It is now accepted that the hydrolysis of acetylcholine to choline and acetic acid is catalysed by an enzyme, cholinesterase in animal system. The enzyme prevents the excessive accumulation of acetylcholine at cholinergic synapse and at the neuromuscular junction. The cholinesterase at a neuromuscular junction is capable of hydrolyzing some 10^{-9} molecules (2.4×10^{-7}) of acetylcholine in one milli second.

The Km in normal heart of Cyprinus carpio is 1.37×10^{-3} M and 1.87×10^{-3} M in Channa punctatus.

It is reported that Km changed to 1.83×10^{-4} M and 2.86×10^{-4} M when the fish was subjected to 4.6×10^{-6} and 2×10^{-4} concentrations of methidathion. Similar increasing trends were reported by Gaur and Gaur & Kumar in the heart of Channa punctatus. It goes to 2.78×10^{-3} M when artificial infarction was produced in the Channa heart.

When normal heart is treated with 2 ppm dimethoate, the Km is increased to 3.30×10^{-3} M and Km further increased to 4.07×10^{-3} M when incised heart is subjected to 2 ppm dimethoate. The constant V_{max} in all experiments indicate that inhibition is competitive in nature.

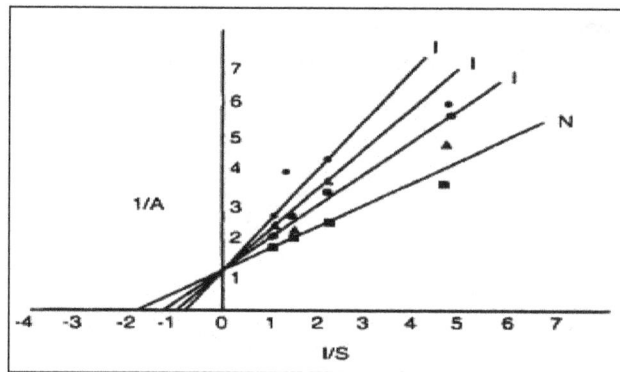

Lineweavers-burk plot of acetylcholinesterase of heart of Channa
punctatus showing inhibition with a pesticide dimethoate, 1/A, absorbance;
1/S, substrate concentration; I, inhibitor; N, normal.

These experiments support that infarction and treatment with pesticide show that in these cases there is inhibition of acetyl-cholinesterase enzyme in the heart tissue.

Endocrine Glands

The glands that secrete their products into the bloodstream and body tissues along with the central nervous system to control and regulate many kinds of body functions are known as endocrine gland. In fishes various endocrine gland has been found associated with different tasks and functions.

The components of endocrine system can be classified on the basis of their organization, which is as follows:

1. Discrete Endocrine Glands: These include pituitary (hypophysis), thyroid and pineal.

2. Organs containing both endocrine and exocrine functions. In fishes, it is kidney, gonads and intestine. Kidney contains heterotopic thyroid follicles, inter-renal, and corpuscles of Stannius.

3. Scattered Cells with Endocrine Function: They are known as diffused neuro-endocrines. They are present in digestive tract. They are generally called as paracrines (e.g., soma-tostatin). There are gastrointestinal peptides whose definite classification as hormone or paracrine agent has not yet been established, these are designated as putative hormones.

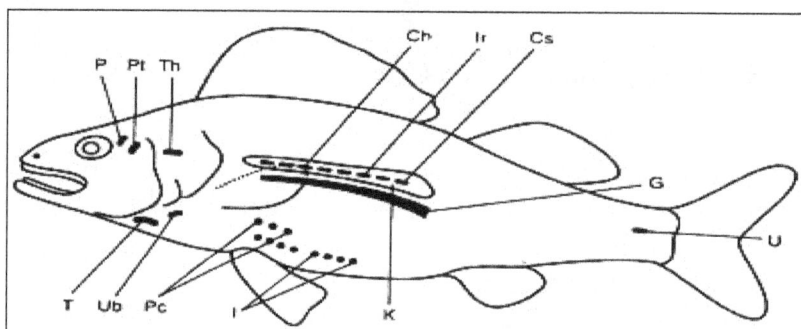

Schematic diagram to show position of various endocrine glands in fishes. Ch. chromaffin tissue;
Cs, corpuscles of Stannius; G, gonad; 1, intestinal tissue; 1r, interrenal tissue; K, kidney; P, pineal; Pc,
pancreatic islets; Pt, pituitary; T, thyroid; Th, thymus; U, urohypophysis; Uls, ultirnobranchial.

Chemically, hormones can be divided into three classes:

1. Steroid hormones (testosterone & estradiol).

2. Protein (peptide) hormones (e.g., insulin) and peptide hormones are secreted by hypophysis, thyroid, internal tissue and pancreatic tissue.

3. The amino acid analogues are norepinephrine and epinephrine, collectively called catecholamines.

Endocrine glands of increasing complexities are found in cyclostomes, elasmobranchs and osteichthyes. Elasmobranchs (sharks) possess well developed endocrine glands but these show some interesting differences from those of higher chordates. However, osteichthyes (bony fishes) have endocrine glands rather more similar to higher chordates.

The difference between fish and mammal endocrine glands is probably due to the development and modification of various body systems in these two classes, and also due to exigencies of an aquatic mode of life.

Mammalian endocrine glands are well advance and well-studied but fish endocrinology is limited to the work on its influence on chromatophores, action of sex cells, function of pituitary and thyroid and control on migration.

Unlike nervous system, the endocrine system is basically related to comparatively slow metabolism of carbohydrate and water by adrenal cortical tissue, nitrogen metabolism by adrenal cortical tissue and thyroid glands and the maturation of sex cells and reproductive behaviour by the pituitary gland and gonadal hormones.

Pituitary Glands

Origin of Pituitary Glands

The pituitary gland occupies the same central part in the endocrine signalling system of fish that it has in mammals. This master endocrine gland originates embryo-logically from the two sources. One as ventral down-growth of a neural element from the diencephalon called the infundibulum to join with another, an ectodermal up-growth (extending as Rathke's pouch) from primitive buccal cavity.

These two outgrowths are thus ectodermal in origin and enclose mesoderm in between them, which later on supply blood to the pituitary gland, originating from the inter-renal carotid artery.

Location of Pituitary Glands

The pituitary gland is located below the diencephalon (hypothalamus), behind the optic chiasma and anterior to saccus vasculosus, and is attached to the diencephalon by a stalk or infundibulum. The stalked pituitary is found in Barbus stigma and Xiphophorus maculatus.

The size of infundibulum varies according to the species. Usually in cyclostomes it is smaller but increases in bony fishes, with prominence in groove or depression of para-sphenoid bone receiving the gland. There is no sella turcica comparable to that found in mammals in Xiphophorus. The short, thick- walled, hollow infundibular stalk contains a lumen, which continues with the third ventricle.

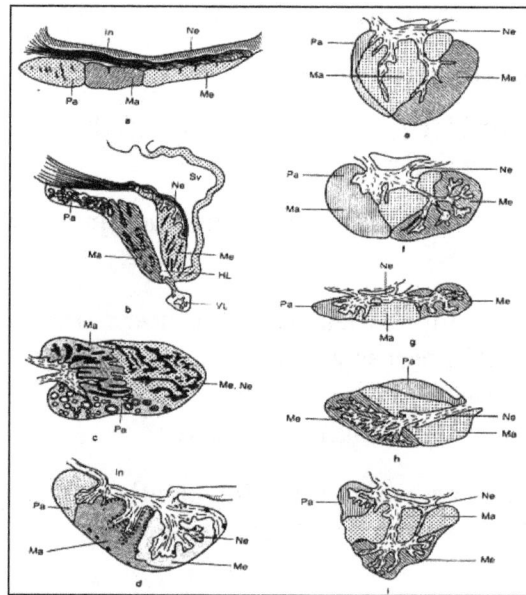

Diagrams of pituitary of various fishes.

In figure, (a) Pdromywn. (b) Dogfish shark (Squalus). (c) Trout (Saline). (d) Perch (Perm). (e) Rainbow trout. (f) Ayu (P. altivelis). (g) Eel. (h) Carp. (i) Yellow tail. HL, lumen of hypophysis; In, Infundlbulum; Ma, mesoadenohypophysis; Me, metaadenohypophysis; Ns, neurohypophysis; Pa, proadenophypophysis; SV, saccus vasculosus; VL, ventral lobe.

Shape and Size of Pituitary Glands

The pituitary is an oval body and is compressed dorsoventrally. The size of sexually mature platy-fish has a mean anterior posterior length of 472.9 micra, with mean width of 178 micra and mean depth of 360 micra. Male glands are smaller than those of females.

On ventral aspect the gland gradually tapers caudally from rounded anterior end. The dorsal surface of the pituitary of platy-fish is concave, ventrally it is slightly convex. The pituitary gland is completely enveloped by a delicate connective tissue capsule.

Anatomy of the Pituitary Glands

Microscopically, the pituitary gland is composed of two parts:

- Adenohypophysis, which is a glandular part originated from the oral, ectoderm.

- Neurohypophysis, which is a nervous part originated from the infundibular region of the brain. Both parts are present in close association.

Pickford and ATZ divided adenohypophysis into three parts, viz., pro-adenohypophysis, mesoadenohypophysis and metaadenohypophysis while Gorbman divided adenohypophysis into three parts but called them as rostral pars distalis, proximal pars distalis and pars intermedia.

However, nomenclature is synonyms as follows:

- Pro-adenohypophysis – Rostral pars distalis.

- Mesoadenohypophysis – Proximal pars distalis.

- Metaadenohypophysis – Pars intermedia.

1. Rostral Pars Distalis (Pro-Adenohypophysis): Lying dorsal to the mesoadenohypophysis in the form of thin strip.

2. Proximal Pars Distalis (Mesoadenohypophysis): Lying almost in between the rostral pars distalis and pars intermedia.

3. Pars Intermedia or Metaadenohypophysis, viz.: Lying at the distal tapering end of the pituitary gland. Pituitaries are broadly characterized as platybasic and leptobasic. In platybasic form (Eel), the neurohypophysis consists of flat floor of the caudal infundibulum which sends processes into disc-shaped adenohypophysis.

In leptobasic, the neurohypophysis has a fairly well developed infundibulum stalk and the adenohypophysis is globular or egg shaped. There are many intermediate between the two.

Adenohypophysis

Earlier workers identified cells of adenohypophysis on the basis of staining procedures. The procedures used were Heidenhain's azon method, Masson's poncean acid fuschin anilin blue, the periodic acid Schiff's reaction (PAS), aldehyde fuschin technique (AF) and then they made cell counts.

The cells of pituitary secrete hormones and hormones are stored in granules present in the cytoplasm. The cells are, therefore, classified on the basis of staining properties of granules of these cells. Cell types of adenohypophysis, on the basis of staining reaction, to the mixture of acidic and basic dyes with secretory granules are called as acidophilic and basophilic.

On the basis of binding affinities with ribonucleoprotein the two classes are also classified as chromophobes and chromophils. The chromophobes have little affinity with dye while chromophils stain strongly as they have affinity with dye.

Chromophilic cells which take acidic stain are called as acidophils whereas the chromophilic cells which bound basic dye are called basophilic and the cells which do not take any stain are called chromophobes. The acidophilic cells are PAS (periodic acid Schiff) and AF (aldehyde fuschin) negative cells. The basophilic cells are AF and PAS positive.

Recently on the basis of immunocytochemistry, the cells are classified according to hormones released by the pro-adenohypophysis.

For example, the cells which take basophilic stain but produce adrenocorticotropic hormones, they are called ACTH cells but if secrete thyroid stimulating hormone, these cells are called thyrotrophs and if they secrete FSH hormones they are called gonadotrops although they are basophilic in nature.

The cells of adenohypophysis when stained with periodic and Schiff (PAS) and aldehyde fuschin methods/if do not take stain, they are PAS and AF negative.

The teleost hypo-thalamo-pituitary system is unique amongst vertebrates, as there is direct inner-vation of pars distalis by neurosecretory neurons of hypothalamus and there is loss of modification of the typical vertebrate hypothalamohypophysial portal vascular system for transport of neuro-hormones to pars distalis.

a. Pro-Adenohypophysis: It contains cells which secrete prolactin and corticotropin (ACTH) exclusively in addition to other hormones.

b. Mesoadenohypophysis: The mesoadenohypophysis (proximal pars distalis) contains cells which produce gonadotropin (GTH) and growth hormone (GH). Thyrotropin cells may oc-cur in either or both in rostral pars distalis and proximal pars distalis. Acidophils rounded or oval or sometime pyrimidal shaped.

    They are coarsely granular, and give the cytoplasm a splotched appearance. They have round to oval peripheral nuclei. Basophils (cyanophilic) are spherical with large, round, centrally located nuclei. Their cytoplasm is finely granular, chromophobes cells are similar in structure as they are found in pro-adenohypophysis. The basophilic (cyanophilic) cells are PAS positive and AF positive.

c. Metaadenohypophysis: It also encompasses more neurohypophysial tissue than any other region. The meta-adenohypophysis basophilic cells are PAS positive. However, the granu-lar cells do not show consistent staining reaction with PAS and AF stains.

Neurohypophysis

The neurohypophysis occupies considerable portion of the gland and possesses many interesting and distinctive features. The neurohypophysis comprises connective tissue, neuroglia cells and loosely tangled network of nerve fibres. These nerve fibres are scattered horizontally along the dorsal part of the adenohypophysis and run vertically, which are generously inter-spread with granular material, large irregularly shaped amorphous masses and large nuclei.

They are located in the mid-dorsal region. The amorphous masses are called "Herring bodies", which have an intimate relation with the di-encephalic neuro-secretory cells called nucleus preop-ticus by means of a fibre tract known as the preoptic neurohypophysial tract.

The diencephalon another parts of the brain contain a group of neurons and each group is called nucleus. The NPO and NLT are important as their axons are in association with both adenohypo-hysis and neurohypophysis . These possess neurosecretory cells.

The nucleus preopticus (NPO, preoptic nucleus) is situated on either side of the optic recessus slightly in front of the optic chiasma. The preoptic nucleus (NPO) is further subdivided into two parts, I, Pars parvocellularis, it is located anteroventrally and consists of relatively small cells, II, pars magnocellularis, it is situated posterodorsally and comprises relatively larger cells.

The preoptic nucleus (nucleus preopticus, NPO), their axons and nerve endings in the pituitary are stainable by neurosecretory stains. The neurons with Gormori's chrome alum haematoxylin, aldehyde fuschin and alcian blue, can differentiate the NPO from other nuclei in preoptic region as they are neurosecretory in nature.

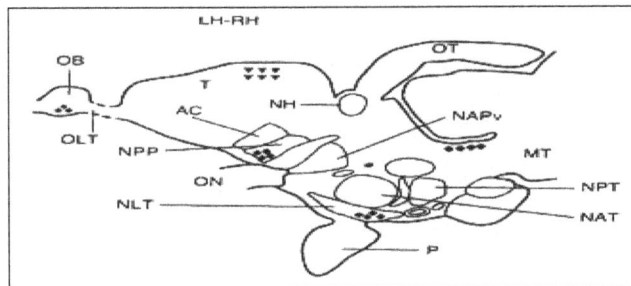

Diagram to show NPO and NLT in the brain of platyfish.

In figure, AC, anterior commissure; MT, mid-brain tegmentum; NAT, nucleus anterioris tuberis; NAPv, nucleus anterioris periventricularis; NH, nucleus habenularis; NLT, nucleus lateral tuberis; NPO, nucleus preopticus; NPP, nucleus prepoticus periven-tricularis; NET, nucleus posterioris tuberis; OB, olfactory bulb OLT, olfactory tract; ON, optic nerve; OT, optic tectum; P, pituitary; T, telencephalon.

Blood Supply in Pituitary Glands

Vascularization of the pituitary has been studied in variety of species. In brook trout, Salvelinus fontinalis and Atlantic salmon, Salmo salar, there is a separate blood supply to the neuro-intermediate lobe from the caudal hypothalamic artery and to the combined rostral proximal pars distalis from hypophysial arteries that branch off the anterior cerebral arteries.

There is no separate blood supply for rostral proximal pars distalis and pars intermedia in Salmo gairdneri but the entire blood supply originated anteriorly from the hypophysial arteries. However, in teleosts, the rostral proximal pars distalis receives blood supply from the extensive looping's of arterioles which are found near the interface with the pars distalis.

These vessels are invaded into the pars distalis together with the interdegitations of the anterior neurohypophysis. It has been considered that these anterior loops are the rudiment of the hypothalamohypophysial portal system. However, there are no neurovascular connections with these blood vessels, as are typically found in the median eminence of various vertebrates. This hypothesis is argued as a portal system.

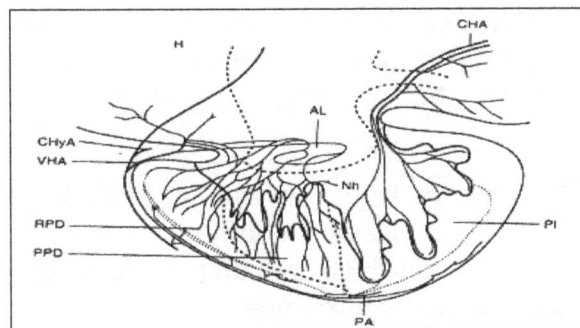

In figure, para sagittal view of pituitary gland of brook trout showing blood supply. AL., arterial loops; CHA, caudal hypothalamic artery; CHyA, caudal hypophysial artery; if, hypothalamus; Nh, neurohypophysis; PA, peripheral artery; PI, pars intennedia; PPD, proximal pars distalis; RPD, rostra) pars distalis; VHA, ventral hypothalamic artery.

The function of the hypothalamohypophysial portal system as a means of transport of neurohormones to the pituitary has become redundant and pure vascular in function, probably because the pituitary cells have direct innervation by neurosecretory endings.

In spite of this, a typical but small hypothalamohypophysial portal system has been described in variety of teleosts. The branches of hypothalamic arteries form "primary capillary plexus" located in the meningeal tissue and the adjacent neural tissue of hypothalamus anterior to the pituitary stalk.

This plexus converges into vessels that enter the pituitary or proximal pars distalis or pars intermedia.

Thus in teleosts this portal system is the only and primary source of blood for the pars distalis. It has been considered that among teleosts the Cypriniformes or Siluriformes have reduced portal system. It is clear, however, that teleosts have neurohormones secreted more or less directly to the pituitary, and that some have the potential for vascular transportation of neurohormones by a portal system as well.

Hormones of Pituitary Glands

There are (seven) various hormones secreted by pituitary but it is generally agreed that one cell type-one hormone concept, is correct. The different hormones secreting cell are not localized in specific region but are spread over in part of the adenohypophysis.

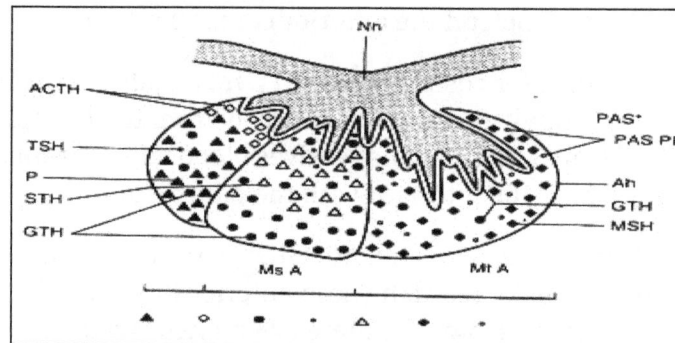

Section of the pituitary to show various hormone
secreting cells in the adenohypophysis (Ah).

Figure shows ACTH, adeno corticotropic cell; GTH, gonado tropic cell; MsA, mesoadenohypophysis MIA, metaadenohypophysis; MSH melanotropic cell; MI, neurohypophysis; PAS, periodic acid Schiff-positive cell in pars intermediate; P prolactin producing cell; STH, somatotropic cell; TSH; thyrotropic cell.

All hormones secreted by the pituitary are necessarily proteins or polypeptides. There exists a slight difference in the pituitary hormones of the different groups of fishes. The pituitary hormones of fishes are of two types (I) one which regulates the function of other endocrine glands. Such hormones are called tropins or tropic hormones.

These are:

1. Thyrotropin activates thyroid.

2. Adrenocorticotropic hormones activate adrenal cortex.

3. The gonadotropin FSH and LH (Leuteotropins, various steroid hormones).

4. Growth hormones, somatotropin (actually they are not tropic).

Second this directly regulates the specific enzymatic reactions in the various body cells or tissues. These hormones are melanin hormones (MH) and melanophore stimulating hormone (MSH), etc. Thyrotropin hormone is secreted from pro-adenohypophysis (rostral pars distalis) and stimulate activity of thyroid hormones.

The TSH is secreted under the influence of (TRH), thyroid- releasing hormones from diencephalon in fishes. It is proved that TRH influences the TSH cell activity and thyroid activity in fish. In Carassus auratus, crude extract of the hypothalamus or goldfish results decreased radioiodine uptake by the thyroid, which indicates the presence of TRH activity in hypothalamus.

In teleosts, the TSH cells have direct innervation by neurosecretory endings, which are adjacent to the cells having no synaptic contact or the endings and may be separated from the TSH cells by a basement membrane. In Tilapia mossambica and Carassius auratus the TSH cells have direct contact with endings containing elementary neurosecretory granules, and with endings containing vesicles having dense granules.

Gonadotropin

Gonadotropin (GTH) cells are richly found in the proximal pars distalis (PPD), where they may form a solid ventral rim of cells. Such situation is found in Cyprinoide. In salmonids and eel they are spread throughout rostral pars distalis (RPD) and PPD.

Gonadotrops are basophilic cell types and are PAS and AB positive. These cells have irregular and more or less dilated cisternae of granulated endoplasmic reticulum (GER) containing granules with varying electron density.

The gonadotroph (GTH) is under the control of gonadotropin releasing hormone. In many teleosts, unlike mammals, neurosecretory stimuli may pass along the nerve fibres piercing the lamina that separates the neuro from adenohypophysis, and penetrating into the endocrine parenchyma of pars distalis. There are two types of nerve fibres designated as A and B types.

The fibres of A type remain in contact with hormone producing cells, including gonadotrops and even terminate with synapse on these cells. B type fibres form synaptic contact with a large granular vesicle of 60-100 nm diameter, while the A synapse have granules of 100-200 nm diameter.

The gonadotropin (GTH) releasing hormone (GnRH) of teleost is similar to luteinizing hormone releasing hormone (LH-RH) is localized in ventral lateral nucleus preopticus periventricularis (NPP) and posterior lateral nucleus lateral tuberis (NLT) as well as other areas.

In hypothalamus, localization of immuno reactive fibre tracts from cells in the NPP and NLP to the pituitary gland suggests that these areas are the origin of endogenous releasing hormone.

Studies of Peter and Crim on Carassius auratus indicate that the nucleus lateralis tuberis (NLT) pars posterior and the NLT pars anterior which are situated in the pituitary stalk, are actively take part in regulation of GTH secretion for gonadal recrudescence.

In several fishes GTH secretion is associated with ovulation. In Carassius auratus, GTH level becomes higher on the day of ovulation. However, in sockeye salmon, Oncorhynchus nerka, high level of GTH found during spawning.

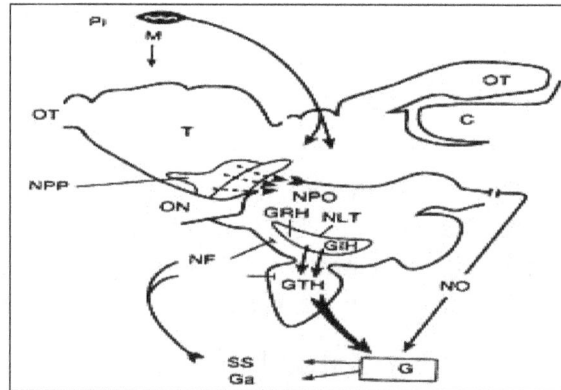

Figure shows diagrammatic representation of the neuroendocrine regulation of gonadotropin (CTH) secretion by releasing hormone (CRH) and one inhibitory hormone (CIH) in gold fish brain. C, cerebellum; C; gonad; Ca, gametes; M. melatonin; NF, negative feedback; NLT, nucleus latcralis tuber's; NPP, nucleus preopticus periventricularis, NPO, nucleus preopticus; NO, neural output; ON, optic nerve; OT, olfactory tract; Pt, photoperiod input; SS, sex steroid; T, telencephalon.

In fishes there is only one functional gonadotropin is found, which is often regarded as piscian pituitary gonadotropin (PPG). This single gonadotropin has similar properties of two hormones. LH and FSH of mammals. Mammalian luteinizing hormone (LH) promotes release of gametes from nearly mature gonads in fishes and stimulates appearance of secondary sexual characters.

This indicates that there must be a similar hormone in fishes also. Salmon pituitary secretes gonadotropins which resembles LH. Furthermore, the gonadotropins from human chorion and urine of gravid mares have LH like properties which hasten the release of eggs in female fishes.

The presence of follicle stimulating hormone in fishes (FSH), which is the second gonadotropin hormone, found in mammalian pituitary gland, is still not confirmed. Recently, prostaglandin, which is hormone-like substance, has been isolated from testis and semen of bluefin tuna, Thynnus thynnus and flounder (Paralichthys olivaceus).

Adrenocorticotropic Hormone (ACTH)

It is secreted by ACTH cells located between the rostral pars distalis and the neurohypophysis. Secretion of ACTH from pituitary is stimulated by the hypothalamus through corticotrophin releasing factor (CRF).

Hypothalamic and telencephalic extracts of Carassius auratus and longnose suckers, Catostomus catostomus stimulated secretion of ACTH in Carassius auratus in vivo. The nature of this telencephalic hypothalamic CRF is unknown. However, it shows similarity with mammalian CRF.

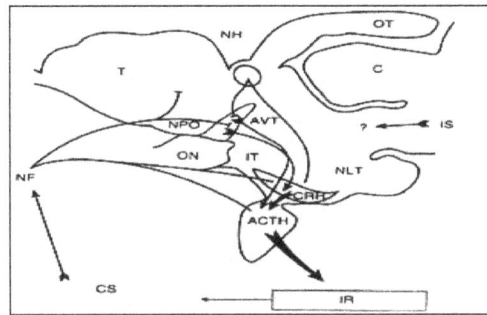

Figure shows diagrammatic illustration of neuroendocrine regulation of (ACM) by a releasing hormone (RH) and (NPO)—) arrow indicates stimulatory influences; Question (?) indicates unknown pathways. AVT, arginine vasotocin; IT, isotonin; IR, inter renal; IS, input of stress; C, cerebellum; CS, cortico steroid; NF, negative feedback; NH, nucleus habenularis, NLT, nucleus lateralis tuberis; NPO, nucleus preopticus; ON, optic nerve; T, telencephalon.

In Carassius auratus, ACTH cells are innervated by aminergic like type B fibres, which originate from nucleus lateralis tuberis (NLT). In teleosts, neurohypophysial peptides may regulate the ACTH secretion.

Implantation of cortisol pellets in Carassius auratus shows that corticosteroids exert negative feedback effects on the brain to suppress ACTH secretion. Cortisol added to the medium inhibits the activity of ACTH cells and release of ACTH which also suggest direct negative feedback effect of Cortisol on the ACTH cells.

Prolactin

It is a similar hormone that influences lactation in mammals and is released from pro-adenohypophysis. In some fishes like mummichog (Fundulus heteroclitus), prolactin along with the intermedin enhances the laying down of melanin in the melanophores of the skin.

Among the several hormones the prolactin is also involved in electrolytic regulation in teleosts but its importance in maintaining homeostasis varies according to species. The secretion of prolactin from the teleost pituitary is under an inhibitory neuroendocrine control of hypothalamic origin.

Growth Hormone (GH)

Mesoadenohypophysis secretes a growth hormone which accelerates increase in the body length of fishes. Very little is known concerning its control, mode of action on cell division and protein synthesis in teleosts. It has been reported that teleost GH cells are capable of some spontaneous activity and continue to synthesize and secrete GH in vitro.

It is evident that the GH secretion may be influenced by osmotic pressure, as release of GH from cultured Salmo gairdneri and Anguilla Anguilla pituitaries greater in a medium containing low sodium than in a high sodium medium, relative to plasma sodium levels. However, GH release from Poecilia latipinna pituitary has no effect of osmotic pressure.

Recently a zone of the hypothalamus has been recognized, which is believed to be responsible for control of GH in Carassius auratus. In this fish nucleus anterior tuberis (NAT) and sometimes nucleus lateralis tuberis (NLT) forms an area which stimulates GH secretion and is perhaps the origin of a growth hormone releasing hormone (GRH).

The hypothalamic control of GH secretion is revealed by ultra-structural studies on the pituitary. In teleosts GH cells of pars distalis have direct synaptoid contact with type B endings as in Carassius auratus, which have direct contact without the synaptoid appearance in Tilapia mossambica.

Very few species like Oryzias latipes contain synaptoid contact of type A endings on GH cells; in other teleosts, type A fibre may have direct contact with GH cells, but generally the endings are separated from the cells by a basement membrane. Thus it is clear that on neuroendocrine factor reaching the GH cells and probably the GH cells are regulated by a dual hormone.

Melanocyte Stimulating Hormone (MSH) or Intermedin

MSH is secreted from the meta-adenohypophysis and acts antagonistically to melanin hormone (MAH). MSH expands the pigment in the chromatophores, thus takes part in adjustment of background. It also stimulates the melanin synthesis. Pars intermedia of teleost pituitary comprise two kinds of secretory cells, which can be identified by their staining properties.

One cell type is PAS^{+ve} periodic acid Schiff positive and PbH^{-ve} (lead hematoxylin negative). However second cell type is PbH^{+ve} and PAS^{-ve}. Salmonid seems to have only PbH^{+ve} cells.

These cells are source of melanocyte stimulating hormone (MSH) which stimulates melanin despersion in the melanocytes and darkening of skin. Neuro-intermediate lobe of Salmo gairdneri appears to have a melanin concentrating factor.

The occurrence of MSH and/or its precursor ACTH in PbH cells of several species of teleosts has been demonstrated by immuno fluorescence techniques. Thus these observations confirm earlier correlations of body colour or background adaptation with activity of the PbH cells.

In teleosts, the neuroendocrine control of pars intermedia varies according to species. In fishes like Cymatogastes aggregate, Anguilla Anguilla and Salmo gairdneri, neurosecretory axons do not enter the pars intermedia but terminate in extravascular channels bordering the pars intermedia or terminate at the basement membrane.

However, other teleosts such as Carassius auratus, and Gillichthys mirabilis have direct innervation from neurosecretory axons.

Secretion of MSH in teleost may be suppressed by catechol aminergic mechanism. Treatment of 6 OHDA to destroy catechol aminergic nerve terminals also causes activation of the MSH cells in Gillichthys mirabilis, darkening of skin or activation of MSH cells in Anguilla Anguilla, catecholamine inhibits the release of MSH directly, when the former is MIH. Also the catecholamine may affect MSH secretion indirectly by promoting the release of an MIH, or by inhibiting the secretion of MRH nerve terminals within the pars intermedia.

Several histological investigations demonstrate a stimulation of the pars intermedia associated with reproduction. During spawning period pars intermedia of clupea becomes intensely active. In Carassius auratus the number of PbH^{+ve} cells increases after spawning in both number and activity during oogenesis and breeding season.

Oxytocoin and Vasopressin Hormones

In fishes the neurohypophysis secretes two hormones, i.e., oxytocin and vasopressin, which are stored in hypothalmic neurosecretory cells. These endocrine substances have well known effect on mammalian metabolism.

Vasopressin and antidiuretic (ADH) hormones are responsible for the constriction of blood vessels in mammals and thus stimulates retention of water by their action in kidney. Oxytocin stimulates mammalian uterine muscles and increases the discharge of milk from lactating mammals.

The fish pituitary hormones are capable to produce such effects in higher vertebrates but presumably the target organ is specific site of their action in fishes and probably is different from those of higher vertebrates. In fishes they control osmoregulation by maintaining water and salt balance.

Use of Pituitary Hormones in Induced Breeding

The pituitary hormones have practical applications by injecting and implanting to force or stimulate spawning of certain fishes of great economic value, such as trouts (Salmoninae), catfishes (Ictaluridae), Mullets (Mugliliade), and sturgeons (Acipenseridae). The synthesis of sex hormones in the gonad is controlled by pituitary gonadotropin.

Hence pituitary extract containing GTH are taken from sexually mature male or female fish than injected to the same species for inducing and hastening and spawning. For the preparation of pituitary extract closely related species may also be used as donor.

Thyroid Gland of Fishes

Location of Thyroid Gland

In many teleosts the thyroid gland is situated in the pharyngeal region in between the dorsal basibranchial cartilages and ventral sternohyoid muscle. The thyroid surrounds anterior and middle parts of first, second and sometimes third afferent branchial arteries of ventral aorta, as found in Ophiocephalus species.

In Heteropneustes it occupies almost the entire length of the ventral aorta and afferent arteries. In Clarias batrachus the thyroid gland is concentrated around the ventral aorta, middle ends of two pairs of afferent arteries and the paired inferior jugular veins.

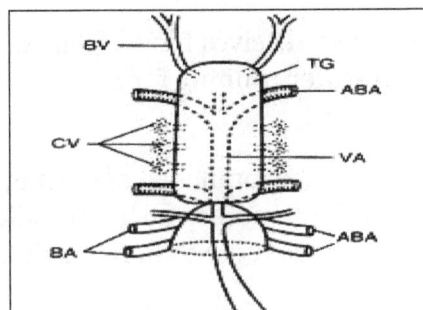

Ventral view of thyroid gland and their blood vessels in Heteropneusles.
ABA, afferent branchial artery; BA, branchial vessel; BV, buccal vein;
CV, commissural vessel; TC, thyroid gland; VA, ventral aorta.

Shape and Size of Thyroid Gland

In majority of teleosts the thyroid is un-encapsulated and thin follicles are dispersed or arranged in clusters around the base of afferent branchial arteries.

It is thin-walled, sac-like, compact dark brownish and enclosed in a thin-walled capsule of connective tissue in these fishes might be correlated with the air breathing habit because thyroid gland acts here as thermoregulatory to adapt the fish to a semiterrestrial environment of low thermal capacity.

In Heteropneustes the thyroid gland is an unpaired thin- walled brownish but cylindrical in shape. In Clarias batrachus the thyroid gland is not covered by definite wall, i.e., un-capsulated and is elongated in shape.

Histology of the Thyroid Gland

In teleosts, histologically the thyroid gland consists of a large number of follicles, lymph sinuses, venules and connective tissues. The follicles are round, oval and irregular in shape. Each follicle contains a central cavity surrounded by a wall composed of single layer of epithelial cells. The structure of epithelium varies according to its secretory activity. Less active follicles generally have thin epithelium.

Epithelial Cells are of two Types

1. Chief cells which are columnar or cuboidal in shape, having oval nuclei and clear cytoplasm.

2. Colloid cells or Benstay's cells. They possess droplets of secretory material. The follicles are supported in position by connective tissue a fibre, which surrounds them. The central lumen of follicle is filled with colloid containing chromophilic and chromo phobic vacuoles.

Blood Supply in Thyroid Gland

The thyroid gland is highly vascularized and is generally well supplied with blood. A single buccal vein and two pairs of commissural vessels supply blood to thyroid gland. From its posterior end a pair of veins arise, which merge immediately to form the posterior inferior jugular vein sending blood to the heart.

In Ophiocephalus, the thyroid gland also receives blood from same vessels. The buccal vein collects blood from the buccal region and after running for a short distance beneath the anterior end of pairs of the thyroid gland opens into it.

The two commissural blood vessels are highly branched and bring blood from the floor of pharynx. One pair opens at the anterior end while the other at the middle of the gland on either side. In Heteropneustes the commissural vessels are more than two pairs.

Hormones of Thyroid Gland

Thyroid hormone is synthesized in the thyroid gland, for which inorganic iodine is extracted from the blood. This inorganic iodine combines with tyrosine. The thyroid hormones of fishes appear to be identical with those of mammals, including, mono- and di-iodo-tyrosine and thyroxin.

These hormones are kept stored in the thyroid follicles and are released into blood stream on metabolic demands. The release of the thyroid hormone from the follicle is controlled by the thyrotropic hormone (TSH) of pituitary which in turn is influenced by genetically determined maturation process along with certain factors like temperature, photoperiod and salinity. The thyroid glands in sharks and higher teleosts are diffused in nature.

Therefore, it is difficult to remove or inactivate. In spite of the certain deficiency, studies have been made by physiological blocking or radio-thyroidectomy using. In teleosts, there is no respiratory stimulation by thyroxin, which is best known in mammals. Physiologically used small quantity of thyroxin and tri-iodo-thyroxine result in thickening of the epidermis and fading of goldfish (Carassius auratus).

Induced thyroid hyperactivity accelerates the transformation into juvenile smolt stage in salmon but high thyroid the titer retards growth of the larva in the same genus. Induced thyroid hyperactivity in mud skipper (Periophthelamus) shows morphological and metabolical changes in response to the more terrestrial existence of fish living mostly outside the water. Thyroid gland of salmons and sticklebacks is known to influence osmoregulation.

In salmon the thyroid gland becomes hyperactive during their spawning migration. It has been considered that thyroid influences the growth and nitrogen metabolism in goldfish, as indicated by high ammonia excreted by them. Thus the action of thyroid is conjugated with other vital processes including growth and maturation and also the diadromous migration of fishes.

Adrenal Cortical Tissue or Inter-renal Tissue

Location

In Lamprey (Cyclostomata) the endocrine inter-renal cells are present throughout the body cavity close to the post-cardinal vein. Among the rays they lie in more or less close association with posterior kidney tissue, including some species possessing inter-renal tissue concentrated near the left and in other near the right central border of that organ.

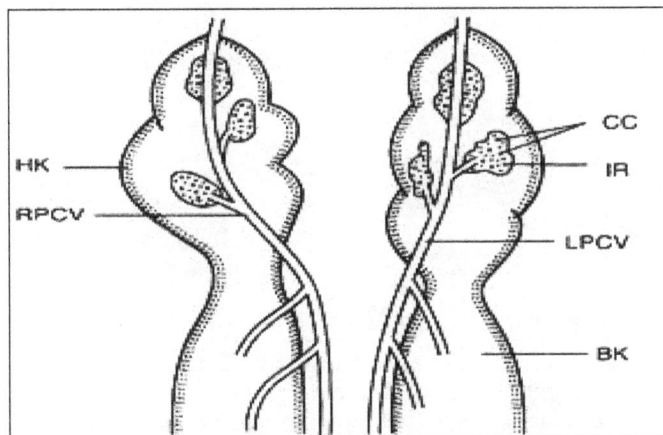

Digram to show location of interrenal glands in fishes. BK, body kidney, CC, chromaffin cells; HK, head kidney; IR, interrenal tissue; LPCV, left post cardinal view; RPCV, right post cardinal vein.

In sharks (Squaliformes) they are present between the kidneys. In teleosts the inter-renal cells are multilayered and situated along the post-cardinal veins as they enter the head kidney.

Anatomy

In some fishes like Puntius ticto inter-renal cells are arranged in form of thick glandular mass while in others like Channa punctatus they are present in form of lobules. Each inter-renal cell is eosinophilic and columnar with a round nucleus.

Adrenal Cortical Hormone

Adrenal cortical tissue or inter-renal tissue secretes two hormones. These are (i) mineral corticoids concerned with fish osmoregulation, (ii) glucocorticoids, which regulate the carbohydrate metabolism, particularly blood sugar level.

Salmo gairdneri treated with mineral corticoid excretes higher than normal amount of sodium ions through its gills but conserve more than normal amount of sodium in the kidneys and osmoregulation in the body.

Intramuscular injection of corticosteroid compounds to the oyster toadfish causes increase in blood sugar level thus showing control on carbohydrate metabolism. The cortisone level of blood plasma of salmons rises during spawning period and declines during the more sedentary stages.

During the spawning phases 60% of total body protein is catabolized in Oncorhynchus, which is correlated with the six fold increase in plasma corticosteroids and rises in liver glycogen.

Like higher vertebrates administration of adrenal cortical hormones stimulates lymphocyte release in Astyanax and antibody release in European perch. Corticosteroids structurally similar to androgens and produce androgens side effects. Secretion of adrenocortical hormones is under control of the adrenocorticotropic hormone (ACTH) of hypophysis.

Chromaffin Tissue or Suprarenal Bodies or Medullary Tissue

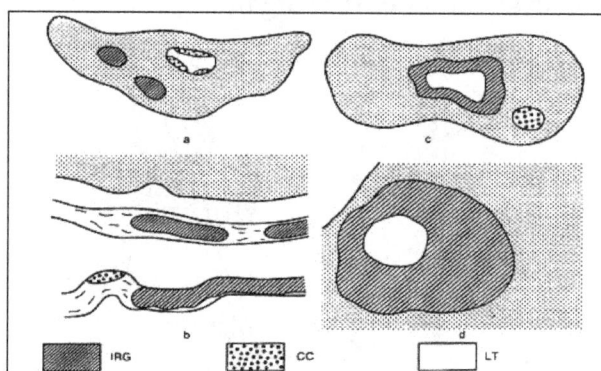

Digram showing chromaffin tissue in fishes
(a) Rainbow trout. (b) Eel. (c) Carp. (d) Yellow tail. IRG,
interremal gland; CC, chromaffin cell, lymphoid tissue.

In lamprey (Cyclostomata) the chromaffin cells are present in the form of strands along the dorsal aorta as in the ventricle and the portal vein heart. In sharks and rays (Elasmobranchii) these tissues are found associated with the sympathetic chain of nerve ganglia while in bony fishes (Actinopterygii) the chromaffin cells have wide variation in their distribution.

They are elasmobranch like, distributed as in flounders (Pleuronectus). On the other hand they have true adrenal arrangement as in sculpins (Cottus) where chromaffin and adrenal cortical tissues are joined into one organ, similar to the mammalian adrenal gland.

Chromaffin tissue of fishes richly contains adrenaline and noradrenaline. Injection of adrenaline and noradrenaline causes changes in blood pressure, bradycardia, branchial vasodilation, diuresis in glomerular teleosts and hyperventilation.

The Ultimo-branchial Gland

Typically the gland is small and paired and is situated in the transverse septum between the abdominal cavity and sinus venosus just ventral to the oesophagus or near the thyroid gland. Embryonically the gland develops from pharyngeal epithelium near the fifth gill arch. In Heteropneustes, the gland measures 0.4 x 1.5 mm in diameter in average adult of 130 to 150 mm body length.

Histologically, it consists of parenchyma, which is solid and composed of cell cords and clumps of polygonal cells covered by capillary network. The gland secretes the hormone calcitonin which regulates calcium metabolism.

Calcitonin is said to be related with the osmoregulation. Eel calcitonin causes decrease serum osmolarity, sodium and chloride in Japanese eels. The ultimo-branchial gland is under the control of pituitary gland.

The Sex Glands as Endocrine Organs

The sex hormones are synthesized and secreted by specialized cells of the ovaries and testis. The releases of sex hormones are under the control of mesoadenohypophysis of pituitary. In fishes these sex hormones are necessary for maturation of gametes and in addition secondary sex characteristics such as breeding tubercles, colouration and the maturation of gonopodia.

In elasmobranch (Raja) and in salmon the blood plasma contains male hormone testosterone with a correlation between plasma level and the reproductive cycle. Oryzias latipes (medaka) and sockey salmon comprise another gonadal steroid, i.e., 11-ketotestosteron, which is 10 fold more physiologically androgenic than testosterone.

Ovary secretes estrogens of which estradiol-17B has been identified in many species in addition of presence of estrone and estriol. In some fishes progesteron is also found but without hormonal function.

There is little information about the influence of gonadial hormones on the reproductive behaviour of fish. Injection of mammalian testosterone and estrone to lamprey causes development of its cloacal lips and coelomic pores, which contribute in reproductive process.

Such tests conducted for rays and sharks (Elasmobranchii) could not give any results whereas ethynil testosterone (pregnenolone) which produces mild androgenic and progesteron like effects in mammals and birds, found to be highly androgenic in fishes.

Male sex hormones are more similar to those of vertebrates than the ovarian hormones; the former strongly influences ovarian development in a loach, the Japanese weather-fish.

Corpuscles of Stannius

The corpuscles of Stannius were first described by Stannius as discrete gland like bodies in the kidney of sturgeon. The corpuscles of Stannius (CS) are found attached or lodged in the kidneys of fishes particularly holostean and teleost.

Corpuscles of Stannius are asymmetrically distributed and often resembles with cysts of parasites but lie different from the latter by higher vascular supply and dull white or pink colour. Histologically, they are similar to the adrenal cortical cells. Their number varies from two to six according to species.

The CS. may be flat, oval as in goldfish, trout, and salmon. It is made up of columnar cells which are covered by a fibrous capsule. The columnar cells are of two types (i) AF-positive and (ii) AF negative. They are filled with secretory granules. The parenchyma of CS comprises vasculo-ganglionic units consisting of a bunch of ganglion cells, blood vessels and nerve fibres.

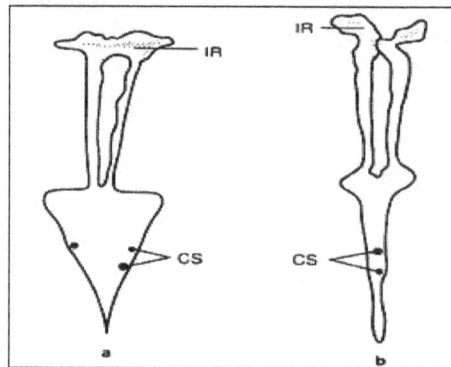

Diagram of kidney of fishes showing corpuscles of Stannius.
(a) Cirrhina mrigala. (b) Labeo rohita. CS, corpuscles
of Stan-nius; IR, interrenal corpuscles.

The number and position of CS vary in different species. There may be single CS as in Heteropneustes setani and Notopterus notopterus, while as many as ten CS are present in some species like Clarias batrachus. In other species their number varies from one to four.There is a gradual reduction in number of CS that has occurred during evolution of Holostei and Teleostei.

In salmonids the CS is located near the middle part of mesonephros but in majority of fishes they are situated at the posterior end of kidney. Garrettpointed out that CS moves progressively backward during the course of evolution as a result of body cavity rather than a migration of CS.

In Notopterus notopterus the CS are present in anterior end of kidney perhaps because most of the body cavity is occupied by the air bladder and also other organs are compactly arranged in limited space.

The presence of the CS at the extreme anterior end of the kidney in Heteropneustes setani is probably because this species has wide body space and long archinephric duct. Therefore, the variation in number and position of CS in teleost species seems to be an embryological speciality.

There is only one-cell type present in the CS of pink salmon. However, two-cell types are found in pacific salmon. The corpuscle of Stannius reduces serum level in the Fundulus heteroclitus, which have environment containing high calcium, such as sea water.

Recently, it has been shown that corpuscles of Stannius work in association with pituitary gland, which exerts hypercalcemic effect, in order to balance relatively constant level of serum calcium.

Intestinal Mucosa

The intestinal mucosa produces secretin and pancreozymin, which are controlled by nervous system and regulate pancreatic secretion. Secretin affects flow of enzyme carrying liquids from the pancreas, whereas pancreozymin accelerates flow of zymogens.

These hormones are usually synthesized in anterior part of the small intestine. In carnivorous fish these hormones are brought into the stomach, containing acidified homogenate of fish flesh or by injection of secretin into gastric vein which stimulates the secretion of pancreas.

Islets of Langerhans

In some fishes like Labeo, Cirrhina, and Channa small islets are present which are separate from pancreas and are found near gall bladder, spleen, pyloric caeca or intestine. Such islets are often referred to as principal islets. But in some species like Clarias batrachus and Heteropneustes fossils the numbers of large and small islets are found to be embedded in the pancreatic tissues, similar to the higher vertebrates.

In fish the islets are big and prominent and consist of three kinds of cells:

1. The beta cells which secrete insulin and take aldehyde fuschin stain,

2. Another type of cells are alpha cells, which do not take aldehyde fuschin stain and have two types, A_1 and A_2 cells, which produce glucagon. The function of the third type of cells is not known. Insulin is secreted by beta cells and regulates the blood sugar level in fishes.

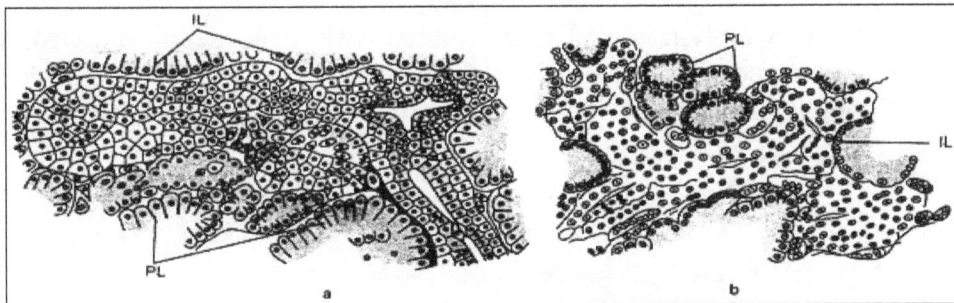

Diagram of pancreas showing endocrine components.
(a) Torpedo marmorean. (b)Mustelus & vols. IL,
islets of langerhans; Pl., pancreatic lobule.

Pineal Organ

It is situated near the pituitary. In-spite of being a photoreceptor organ the pineal organ shows endocrine nature of doubtful function. Removal of pineal from Lebistes species causes reduced growth rate, anomalies in the skeleton, pituitary, thyroid and corpuscles of Stannius. It has been reported that thyroid and pituitary glands influence the secretion of pineal.

Urophysis

Urophysis is a small oval body, present in the terminal part of spinal cord. It is an organ deposits, which releases materials produced in the neurosecretory cells situated in the spinal cord.

These cells together with the urophysis are called the caudal neurosecretory system. This neurosecretory system is found only in elasmobranchs and teleosts but it corresponds to the hypo- thalamo neurosecretory system present in vertebrates.

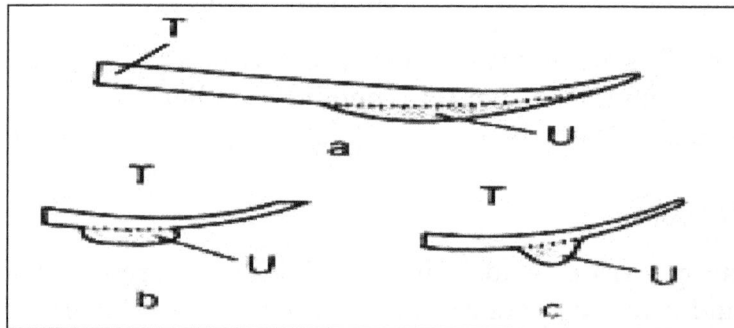

Diagram of urohypophysis of teleosts. (a) rainbow trout.
(b) carp. (c) Yellow tail. T, tail; U, urohypophysis.

In caudal neurosecretory system, neurosecretory cells are diffused in terminal part of spinal cord. Axon terminals of these cells assemble at the ventral side of the region and form urophysis with blood capillaries. The neurosecretory cell is a large nerve cell and has basophilic cytoplasm and a polymorphic nucleus.

In Ayu (P. altivelis) the urophysis is extended like a bow. In carp and yellow tail it is a conspicuous oval body. The urophysis is made up of spinal cord elements like neurosecretory axon, glia, and ependymal and glia fibers and meningeal derivative such as vascular reticulum and reticular fibres.

The caudal neurosecretory system is said to be related with osmoregulation. Urophysis extract shows ability to contract smooth muscles of ovary and oviduct of guppy (Poecilia reticulata) and the sperm duct of goby (Gillichthys mirabilis), suggesting the possibility of involvement in reproduction and spawning.

References

- Fish-anatomy: animalcorner.co.uk, Retrieved 9 August, 2019
- Fish-skeletal-system, biology: byjus.com, Retrieved 7 January, 2019
- Fish-fins-its-types-and-functions: biologyeducare.com, Retrieved 27 April, 2019
- Skin-and-scales-of-fishes, anatomy-and-physiology, fish: yourarticlelibrary.com, Retrieved 9 March, 2019
- Gills, fish, animal: basicbiology.net, Retrieved 29 July, 2019
- Structure-of-gills-in-fishes, anatomy-and-physiology, fish: yourarticlelibrary.com, Retrieved 30 January, 2019
- Structure-fishs-eye-10517: animals.mom.me, Retrieved 14 March, 2019

- Sensory-organs-of-fishes, anatomy-and-physiology, fish: yourarticlelibrary.com, Retrieved 10 June, 2019

- Lateral-line-system-analysis-8971, biology: ukessays.com, Retrieved 28 February, 2019

- Luminous-organs-or-photophore-of-the-fishes, anatomy-and-physiology, fish: yourarticlelibrary.com, Retrieved 1 May, 2019

- Swim-bladder-development-structure-and-types-fishes, fish, fisheries: biologydiscussion.com, Retrieved 2 April, 2019

- Histology-of-gut-in-fishes, anatomy-and-physiology, fish: yourarticlelibrary.com, Retrieved 29 June, 2019

- Heart-structure-of-fishes-structure-pathology-and-innervation, anatomy-and-physiology, fish: yourarticlelibrary.com, Retrieved 22 February, 2019

- Endocrine-glands-of-fishes-fish-anatomy, anatomy-and-physiology, fish: yourarticlelibrary.com, Retrieved 25 May, 2019

Chapter 4

Fish Physiology

The scientific study which focuses on how the different components of fish function together is called fish physiology. Some of the processes studied within this field are respiration, blood circulation, digestion, reproduction and growth. These diverse processes related to fish physiology have been thoroughly discussed in this chapter.

Respiration

Respiration in fish or in that of any organism that lives in the water is very different from that of human beings. Organisms like fish, which live in water, need oxygen to breathe so that their cells can maintain their living state. To perform their respiratory function, fish have specialized organs that help them inhale oxygen dissolved in water.

The respiratory systems of fishes have to be adapted to two important limitations of underwater life. On the one hand, the amount of dissolved oxygen is smaller in the water than in the air: at 23°C, air has 210 ml of oxygen per litre of air, while in freshwater is about 6,6 ml/l and in salt water is 5,3 ml/l. On the other hand, water is much more dense and viscous than air. These limitations explain the adaptations in the breathing of this group of animals.

Respiration in fish takes with the help of gills. Most fish possess gills on either side of their head. Gills are tissues made up of feathery structures called gill filaments that provide a large surface area for gas exchange. A large surface area is crucial for gas exchange in aquatic organisms as water contains very little amount of dissolved oxygen. The filaments in fish gills are arranged in rows in the gill arch. Each filament contains lamellae, which are discs supplied with capillaries. Blood enters and leaves the gills through these small blood vessels. Although gills in fish occupy only a small section of their body, the immense respiratory surface created by the filaments provides the whole organism with an efficient gas exchange.

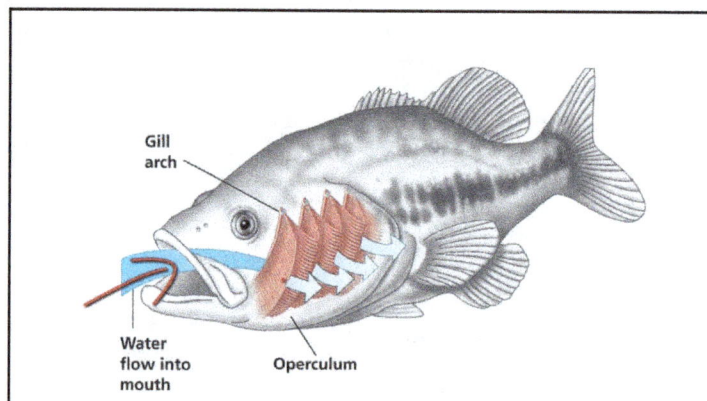

Fish take in oxygen-rich water through their mouths and pump it over their gills. As water passes over the gill filaments, blood inside the capillary network picks up the dissolved oxygen. The circulatory system then transports the oxygen to all body tissues and ultimately to the cells. While picking up carbon dioxide, which is removed from the body through the gills. After the water flows through the gills, it exits the body of the fish through the openings in the sides of the throat or through the operculum, a flap, usually found in bony fish, that covers and protects the fish gills.

Some fish, like sharks and lampreys, possess multiple gill openings. However, bony fish like Rohu, have a single gill opening on each side.

Respiratory mechanism in bony fishes: In bony fishes the main respiratory organ (gill) is covered by the operculum. In all bony fishes water is propelled over the gills by suction pressure. The whole respiratory mechanism of fishes are described by some consecutive steps in below-

- Step 1: At the beginning of respiration, operculum is closed frequently and mouth is opened by the action of sternohyoid and elevator muscle of palatine. At the same time, the brancchiostegal rays are spread and lowered and the mouth cavity enlarged to create negative water pressure in it. Water is thus drawn/enter into the mouth.

- Step 2: After a slight time lag, the space between the gills and the operculum is enlarged as the gill covers are abducted anteriorly. Though the opercular skin flaps are still closed posteriorly by the outside water pressure. The pressure deficit in the gill cavity and the water flows over the gill.

- Step 3: The buccal and opercular cavities begin to reduce while the oral valves prevent the flow of water out of the mouth and the mouth cavity begins to function as a pressure pump instead of suction pump.

- Step 4: In this stage, closed operculum with the opercular flaps has reached its furthest state of abduction and water is accumulating outside the gills. At this point the opercula are quickly brought towards the body, the gill flaps open and the water is expelled, being prevented from flowing backwards by excess pressure in the buccal-cavity as compared to the epibranchial cavity.

Role of blood as a carrier of O_2: Generally O_2 diffuses very slowly from one liquid into another. Thus fishes, like other vertebrates, have evolved in their red blood cells gas-carrying device of high efficiency. Hemoglobin of blood is a respiratory pigment of fishes and other vertebrates found in the red blood cells (R.B.C). Hemoglobin consists of two parts. One is haem i.e. iron-porphyrin part and other is globin i.e. polypeptide. Here the iron atom is Fe^{2+} (Ferrus) form and it converse in Fe^{3+} (ferric) form after combining with oxygen (O_2). Each haem group is enfolded in one of two or four polypeptides called globin.

So it can be said that, haemoglobin is the main functional unit of blood for exchanging or transporting O_2. Hemoglobin acts in two basic ways- i) Giving off O_2 when its tension is lowest in tissue and ii) Binding with O_2 when its tension is higher in gill.

In as much as O_2 taken up from the gill by blood and transported over the body is known as loading blood and when blood returns into gill with a lower density of O_2 is known as unloading blood. It creates much tension/pressure to saturate the blood by O_2 when the amount of CO_2 decreases in blood. On the contrary comparatively lower tension is needed to unloading the O_2. This is known as Bohr Effect.

Role of blood as carrier of CO_2: Generally CO_2 diffuses rapidly in water than O_2 as a result low amount of free carbon-di-oxide in natural water favors waste gas elimination at the gill by diffusion. CO_2 in the venous blood of fishes is carried in solution in the plasma (about 10 poercent), as bicarbonates (about 70%) and as carbamino compound in R.B.C (about 20%). The bicarbonates turn into CO_2 and H2O water by the catalyzation of carbonic anhydrase enzyme. This enzyme is found in acidophil cells of the gills and in other tissues.

CO_2 from the tissues passes to the blood enters into the R.B.C. Then CO_2 turns to carbonic acid and then bi-carbonate and hydrogen ion. The reaction between CO_2 and water normally occurs very slowly but in the R.B.C, the reaction is catalyzed by the presence carbonic anhydrase. Associated chemical reactions are given in below:

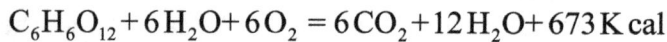

$$C_6H_6O_{12} + 6H_2O + 6O_2 = 6CO_2 + 12H_2O + 673\,K\,cal$$

The reactions of produced CO_2 are here in below:

$$CO_2 + H_2O = H_2CO_3\,(Carbonic\,acid)$$

This H_2CO_3 (Carbonic acid) again breaks down into CO_2 by the reaction of carbonic anhydrase and this CO_2 is released by the respiration of organism or fish.

Accessory Respiratory Organs in Fishes

The different types of accessory respiratory organs in fishes are:

Suprabrachial Organ

The supra-branchial organ is a specialised type of respiratory structure encountered in Clarias batrachus.

It has a complex structural organisation and consists of the following portions:

a.   An elaborate tree-like structure growing from the upper end of the second and fourth gill-arches of either side. This dendritic organ is composed of numerous terminal knobs; each has a core of cartilage covered by vascular membrane. Each exhibits eight folds which suggest that one such knob is formed by the coalescence of eight gill-filaments.

b.   There are pair of highly vascularized supra- branchial chambers within which the tree-like structures are contained. The supra-branchial chambers are developed as the vascularized diverticula of the branchial chamber.

c.   The entrance of the supra-branchial chamber is guarded by 'fan'-like structures which are developed by the fusion of the adjacent gill- filaments of the dorsal side of the gill-arches.

The supra-branchial organs, like the gills, are lined by thin outer epithelial layers with intercellular spaces separated by the pilaster cells. The organs and the supra-branchial chambers are supplied by afferent and efferent blood vessels from the gill-arches.

The supra-branchial organs help to breathe in air. The supra-branchial chamber has inhalant and exhalant apertures. These fishes come to the surface of the water and gulp air into the supra-branchial organs. Atmospheric air from the pharyngeal cavity is taken into the supra-branchial chamber by an inhalant aperture located between the second and third gill- arches.

After gaseous exchange the air from the said chamber expels into the opercular cavity by the gill-slit lying between the third and fourth gill-arches. The fan-like structures present in the second and the third gill-arches help to intake the air while the expulsion of the air from the supra-branchial, chamber is caused by the contraction of its wall. Thus the supra-branchial chamber and its contained organs function as 'lung'.

Branchial Outgrowths

In climbing perch (Anabas testudineus) there are two spacious sac-like outgrowths from the dorsal side of the branchial chambers. The epithelium lining these outgrowths is highly vascular and becomes folded to increase the respiratory area.

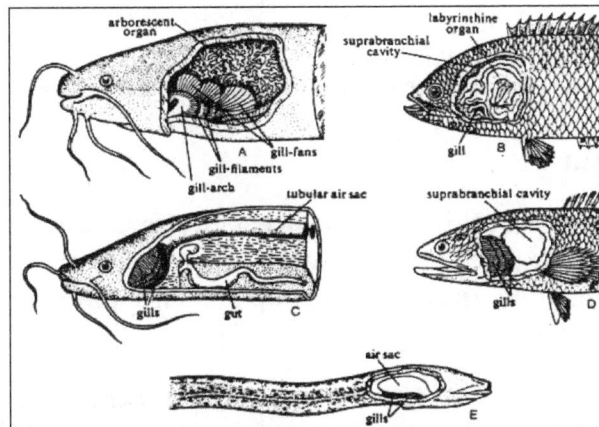

Accessory respiratory organs in air-breadth teleosts. A. Clarias batrachus. B. Anabas testudineus. C. Heteropneustes fossils. D. Channa puntatus. E. Amphipnous cuchia.

Each chamber contains a characteristic rosette-like labyrinthine organ. This organ develops from the first epibranchial bone and consists of a number of shells like concentric plates. The margins of the plates are wavy and the plates are covered with vascular gill-like epithelium.

Each branchial outgrowth communicates freely not only with the opercular cavity but also with the buccopharyngeal cavity. Air enters into the outgrowth by way of the buccopharyngeal opening and goes out through the external gill-slits. The entrance is controlled by valves.

Anabas can breathe in air by the help of these organs. These fishes have the habit of migration from one pond to the other. Their overland progression is peculiar and is assisted by the operculum and the fins. Each operculum bears sharp spines at the free edge.

During travelling the opercula alternately spread out and fix to the ground by the spines and get the forward push from the pectoral fins and the tail. The proverb that the fish can climb the trees seems to be erroneous. The climbing perches are found in the branches of palm or other trees which are possibly brought there by the kites or crows while these fishes migrate over the land.

In Trichogaster fasciatus the accessory respiratory organs are similar to that of Anabas and consist of supra-branchial chamber, labyrinthine organ and respiratory membrane.

The labyrinthine organ is simpler in construction in comparison to that of Anabas. Each organ assumes a spiral configuration with two leaf-like expansions. Each of these two expansions is composed of loose connective tissue which is covered by highly vascular epithelium.

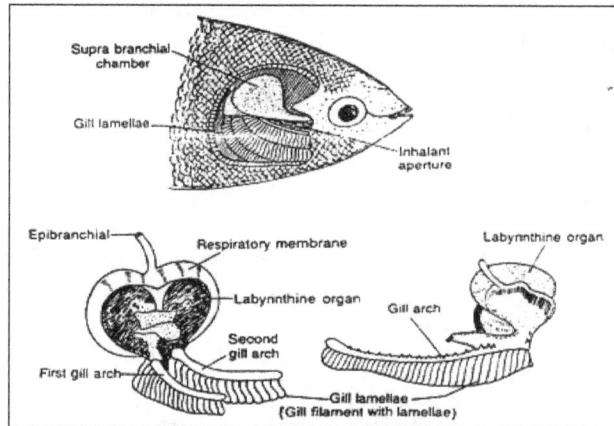

Gills and accessory respiratory organs of trichogaster fasciatus.

Pharyngeal Diverticula

In the Snake-headed fishes and Cuchia eels, the accessory respiratory organs are relatively simplified. These fishes can survive prolonged drought and their air breathing habit enables them to remain out of water for some time. In both the group of fishes, the pharynx gives a pair of saclike diverticula for gaseous exchange.

In Channa, the accessory respiratory organs are relatively simpler and consist of a pair of air-chambers. These are developed from the pharynx and not from the branchial chamber as seen in others.

The air-chambers are lined by thickened epithelium which is highly vascularized. The air-chambers are simple sac-like structures and do not contain any structure. These chambers function as the lung-like reservoirs. In Channa striatus the vascular epithelium lining the chambers becomes folded to form some alveoli. The gill-filaments are greatly reduced in size.

In Cuchia (Amphipnous cuchia) the accessory respiratory organs consist of a pair of vascular sac-like diverticula from the pharynx above the gills. These diverticula open anteriorly into the first gill-slit. These diverticula function physiologically as the lungs.

The gills are greatly reduced and a few rudimentary gill-filaments are present on the second of the three remaining gill- arches. The third gill-arch is found to bear fleshy vascular epithelium. In Periophthalmus, a pair of very small pharyngeal diverticula is present which are lined by vascular epithelium.

Pneumatic Sacs

In Heteropneustes fossil is a pair of tubular pneumatic sacs, one on each side of the body, act as the accessory respiratory organs.

These long tubular sacs arise as the outgrowths from the branchial chamber and extend almost up to the tail between the body musculature near the vertebral column. In Sacco-branchus, similar tubular lung-like outgrowths of the branchial chamber extend back into the body musculature.

Buccopharyngeal Epithelium

The vascular membrane of buccopharyngeal region in almost all the fishes helps in absorbing oxygen from water. But in mudskippers (Periophthalmus and Boleophthalmus) the highly vascularized buccopharyngeal epithelium helps in absorbing oxygen directly from the atmosphere.

These tropical fishes leave water and spend most of the time skipping or 'walking' about through dampy areas particularly round the roots of the mangrove trees. The old idea that the mud-skippers use the vascular tail as the respiratory organ is not supported by recent Icthyologists.

Integument

Eels are recorded to make considerable journey through damp vegetation. The common eel, Anguilla Anguilla can respire through the integument both in air and in water. In Amphipnous cuchia and mud- skippers, the moist skin sub-serves respiration.

Many embryos and larvae of fishes respire through the skin before the emergence of the gills. The median fin fold of many larval fishes is supplied with numerous blood vessels and helps in breathing. The highly vascular opercular fold of Sturgeon and many Catfishes serves as the accessory respiratory structure.

Gut Epithelium

The inner epithelium of the gut essentially helps in digestive process. But in many fishes the gut becomes modified to sub-serve respiratory function. Cobitis (giant loach of Europe) comes above the water-level and swallows a certain volume of air which passes back along the stomach and intestine. In Misgurus fossilis, a bulge just behind the stomach is produced which is lined by fine blood vessels.

The bulge acts as the reservoir of air and functions as the accessory respiratory organ. After the gaseous exchange, the gas is voided through the anus. In certain other fishes, Callichthyes, Hypostomus and Doras the highly vascular rectum acts as the respiratory organ by sucking in and giving out water through the anus alternately.

In these fishes the wall of the gut becomes modified. The wall becomes thin due to the reduction of the muscular layers.

Swim-bladder Acts as Lung

Swim-bladder is essentially a hydrostatic organ but in some fishes it functions as the 'lung'. In Amia and Lepisosteus, the wall of the swim-bladder is sacculated and resembles lung. In Polypterus the swim-bladder is more lung-like and gets a pair of pulmonary arteries arising from the last pair of epibranchial arteries.

The swim-bladder in dipnoans resembles strikingly the tetra- pod lung in structure as well as in function. In Neoceratodus, it is single, but in Protopterus and Lepidosiren it is bilobed. The inner surface of the 'lung' is increased by spongy alveolar structures. In these fishes, the 'lung' is mainly respiratory in function during aestivation because the gills become useless during this period. Like that of Polypterus, the 'lung' in dipnoans gets the pulmonary arteries from the last epibranchial arteries.

In Notopterus, the swim-bladder becomes more complex and acts as a lung. Except the hydrostatic, sound production and hearing, a new function like respiration was innovated in Notopterus. In Notopterus chitala the posterior tip of swim-bladder is enlarged which is called caudal extension and the ventral part gives off several finger-like projections, the dorsal side of the gas bladder possesses a specialised striated muscle.

The anterior part extends into a projection to the ear. An artery arising from the dorsal aorta forms a network of blood capillaries that spread the entire inner surface of the abdominal and caecal parts of the swim bladder.

The blood capillaries that cover a single epithelial layer help in the gaseous exchange between the blood and the air of the swim-bladder. This air breathing habit is considered as a secondary adaptation in these fishes.

Functions of Accessory Respiratory Organs

The accessory respiratory organs contain a high percentage of oxygen. The fishes possessing such respiratory organs are capable of living in water where oxygen concentration is very low. Under this condition these fishes come to the surface of water to gulp in air for transmission to the accessory respiratory organs.

If these fishes are prevented from coming to the surface, they will die due to asphyxiation for want of oxygen. So the acquisition of accessory respiratory organs in fishes is an adaptive feature.

Further it has been observed that the rate of absorption of oxygen in such organs is much higher than the rate of elimination of carbon-dioxide. Hence, it is natural that the gills excrete most of the carbon-dioxide. Absorption of oxygen appears to be the primary function of the accessory respiratory organs.

Significance of Accessory Respiratory Organs

The cause of emergence of the accessory respiratory structures in fishes in addition to the primary respiratory organ is very difficult to interpret. There are two contrasting views regarding the origin of the aerial accessory respiratory structures. First view: some fishes have the natural instinct to make short excursion to the land from the primal aquatic home.

To remain out of water, the development of certain devices to breathe in air becomes necessary. Second view holds that the fishes are forced to ascend the land when the oxygen content of water falls to a considerable extent. The fishes in that particular condition of life gulp in atmospheric air from the land and pass it into the accessory respiratory structures.

If they are prevented by mechanical barriers to come to surface, the fishes will die of suffocation. This habit of swallowing bubbles of air is observed in many bony fishes, especially living in shallow water which dries up periodically or becomes foul by the decomposition of aquatic vegetation.

As a consequence of the air-breathing habit for a considerable span of time, the fishes have developed specialised accessory respiratory organs in addition to the gills.

Most of such structures encountered in the fishes assume the shape of reservoir of air and originate either from the pharyngeal or branchial cavities. In extreme cases the reservoir may house special structure for gaseous exchange.

However, the development of such accessory respiratory organs is essentially adaptive in nature to meet the respiratory need and thus enables the fishes to tolerate oxygen depletion in water or to live on land over a varying period of time. The development of the accessory respiratory organs depends directly on the ability to remain out of the water.

Blood Circulation

The blood of fishes is similar to that of any other vertebrate. It consists of plasma and cellular (blood cells) components. The cellular components are red blood cells (RBC), white blood cells (WBC) and thrombocytes designated as formed elements.

The plasma is liquid portion and consists of water. It acts as the solvent for a variety of solutes including proteins, dissolved gases, electrolytes, nutrients, waste material and regulatory substances. Lymph is the part of plasma that perfuse out of the capillaries to bathe the tissue.

Composition of Blood Plasma

The plasma composition is as follows:

- Water
- Proteins (fibrinogen, globulin, albumin)
- Other solutes
    - Small electrolytes (Na^+ K^+, Ca^{++}, Mg^{++}, CI^- HCO_3^-, $PO+^{--}$, $SO+^{--}$).
    - Non-protein nitrogen (NPN) substance (urea, uric acid, creatine, creatinine, ammonium salts).
    - Nutrients (glucose, lipid, amino acid).
    - Blood gases (oxygen, carbon dioxide, nitrogen).
    - Regulatory substances (hormones, enzymes).

Plasma and Serum

If the blood is prevented from clotting, it separates out into cells and plasma, if it is allowed to clot it separates into clot and serum. Serum and plasma are very similar, the only difference being that serum has lost the clotting factors prothrombin and fibrinogen which are present in plasma.

If the bloods is collected in a vial containing an anticoagulant, the blood will not coagulate and if centrifuged, the blood cells will be separated and settle down, the liquid portion is known as 'plasma'. If the blood is collected in the vial without any anticoagulant, then the blood will coagulate and if this is centrifuged, then the liquid portion is known as 'serum'.

Actually the serum has lost the clotting factor prothrombin and fibrinogen but the plasma contains clotting factor proteins also. The marine teleosts living in polar and sub-polar regions contain antifreeze proteins (AFP) or antifreeze glycoprotein (AFGP). They lower the freezing point of plasma without affecting the melting point.

The fishes which live at low temperature as low as − 1.9 °C do not freeze due to glycoprotein containing amino acids alanine and threonine in the ratio of 2: 1 with molecular weight between 2600 to 33000.

Fish plasma contains albumin, the protein which control osmotic pressure. It also contains lipoprotein whose main function is to transport lipid. Ceruloplasmin, fibrinogen and iodurophorine are some important proteins of fish blood. Ceruloplasmin is a copper binding protein.

The total plasma protein in fish ranges from 2 to 8 g dl^{-1}. The thyroid binding proteins such as T_3 and T_4 are present in the blood circulation in free form. Thyroxine binds to vitellogenin in several Cyprinid species. Enzyme such as CPK, alkaline phosphatase (Alk PTase), SGOT, SGPT, LDH and their isoenzymes are reported in fish plasma.

Formed Elements of the Blood in Fishes

There are three varieties of cells or corpuscles present in blood:

- Erythrocytes
- Leucocytes
  - Agranulocytes
    - Lymphocytes
    - Monocytes
    - Macrophage
  - Granulocytes
    - Neutrophils
    - Eosinophils
    - Basophils
- Platelets or Thrombocytes

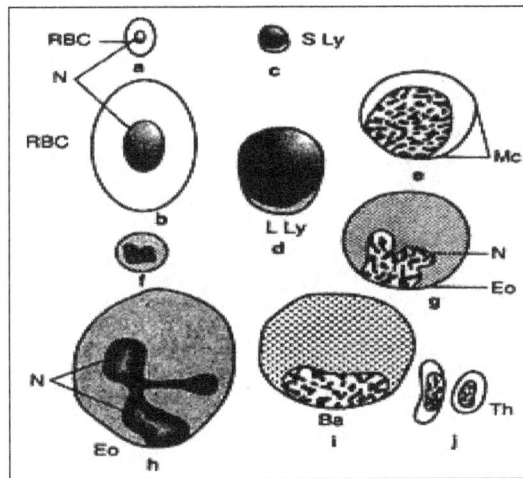

In Figure, Diagrammatic sketch of blood cells of fishes. (a) Erythrocyte (RBC) of elasmobranch (Small). (b) Erythrocyte (RBC) teleost (smaller). (c) Lymphocyte (small) (SLY). (d) Lymphocyte (large) (LLy). (e) Monocyte (Mc). (f) Neutrophll (g) Eosinophil (Eo). (h) Eosinophil (Eo). (i) Basophil (Ba). (j) Thrombocyte (M). N, nucleus; RBC, red blood cells.

Erythrocytes

Dawson classified immature erythrocytes into five categories according to structure, distribution and quantity of basophilic substances within the cell. The cytoplasm of RBC is coppery purplish, coppery pinkish or light bluish in freshwater teleosts. Mature erythrocytes contain abundant haemoglobin and are pink or yellowish on preparation stained with Giemsa.

During the growth of RBC firstly the cytoplasm shows strong basophilia because of the presence of polyribosomes. Due to the accumulation of ultra-cellular protein, the change in the staining reaction of cytoplasm occurs because of haemoglobin which stains with eosin.

The cytoplasm takes eosin stain due to haemoglobin and basophilia due to staining of the ribosomes. Because of double staining the cell is called a polychromatophilic erythroblast. These polychromatophilic erythroblasts are often called reticulocytes.

In adult teleosts, the blood normally contains a certain percentage of immature red cells or senescent cells (growing cells) called pro-erythrocytes or reticulocytes. The number of erythrocyte in blood varies according to the species as well as the age of the individual, season and environmental conditions. However, under similar conditions, a fairly constant number of reticulocytes are present in species.

The nucleus is centrally placed and round or oblong in shape. The RBC size is larger in elasmobranchs in comparison to teleosts. The brackish water species of Fundulus have smaller blood cell than freshwater species. erythrocytes are slightly smaller in active species than in non-active.

In deep sea teleosts, the size of the RBC is larger than normal teleosts. The mature erythrocytes of fishes vary greatly in their shape and outline, and in peripheral blood they are mostly mature. The shape is generally circular in Clarias batrachus, Notopterus notopterus, Colisa fasciatus, Tor tor but ellipsoid, oval or oblong in Labeo rohita and Labeo calbasu.

Abnormal forms of RBC designated as poikilocytes, microcytes, macrocytes, karyorrhexis, basophilic punctuation and nucleated forms were also reported. However, non-nucleated red cells (erythroplastids haemoglobin packets) are formed in Maurolicus milleri, Valencienellus tripunctatus and Vincignerria species in fish blood.

White Blood Cells or Leucocytes

The study of different types of blood cells is done by making a smear on the slide. A drop of blood is placed on the slide and spread thin with another slide. The slide is stained generally with Leishman, Wright or Giemsa stains, often supravital stains like brilliant cresyle blue and neutral red are also used. The stain contains methylene blue (a basic dye), related azures (also basic dyes) and eosin (an acid dye).

The basic dye stains nuclei, granules of basophils and RNA of the cytoplasm whereas the acid dye stains granules of the eosinophils. The basic dyes are metachromatic; they impart a violet to red colour to the material they stain. It was originally thought that neutral dye is formed by the combination of methylene blue and its related azures with eosin, which stains neutrophil granules is not clear.

Although the fish white blood corpuscles have been well investigated, there is no unanimity regarding their classification. The fish leucocytes in peripheral blood are generally (i) Agranulocytes (ii) Granulocytes. The nomenclature is based on affinity of acid and basic dyes and depends upon human haematology. Plasma cells, basket and nuclear shades are also present.

Agranulocytes

They have no granules in the cytoplasm. The most important distinguishing character is un-lobed nuclei. Thus they are distinguished from granulocytes, which possess specific segmented nucleus.

Agranulocytes have two varieties:

    a. Lymphocytes, large and small

    b. Monocytes

Lymphocytes

They are most numerous types of leucocytes. The nucleus is round or oval in shape. They constitute 70 to 90% of the total leucocytes. They are rich in chromatin, although its structure is obscure, and is deep reddish violet in colour in preparation with Giemsa.

The teleost lymphocytes measure 4.5 and 8.2. Also, large and small lymphocytes in peripheral blood smears of teleosts, fresh and marine fishes similar to that of mammals have been noticed. The cytoplasm is devoid of granules but cytoplasmic granules are occasionally present.

In large lymphocytes there is large amount of cytoplasm but in small only little quantity of cytoplasm is evident and nucleus constitutes most of the cellular volume. In fresh samples wandering activity of lymphocytes is rather rare, but leaf like pseudopods may occasionally be seem protruding from the cell.

Functions of Lymphocyte

The main function of fish lymphocytes is to produce immune mechanism by the production of antibody. The T and B lymphocytes are presented similar to T and B lymphocytes of mammals. Teleostean lymphocytes respond to nitrogen, such as PHA, concanvalin A (Con. A) and LPS which are considered as specific to mammalian subclass of lymphocytes.

Klontz demonstrated antibody forming cell in the kidney of rainbow trout while Chiller et al., found antibody forming cells in the head kidney (Pronephros) and spleen of Salmo gairdneri. Plasma cells which synthesize and secrete antibody and immunoglobins are reported in fishes both in light and electron microscopy.

In mammals, when B lymphocytes are activated by an antigen then they transform into immuno-blasts (plasmoblast) which proliferate and then differentiate into, plasma and memory cells.

Monocytes

It consists of much less proportion of WBC population often absent in few fishes. It is suggested that they originate in the kidney and become apparent in the blood when foreign substances are present into the tissue or blood stream. The cytoplasm usually stains smoky bluish or pinkish purple. The nucleus of monocyte is fairly large and varied in shape. The function of monocyte is phagocytic.

Macrophage

They are of large size; the cytoplasm is occasionally finely or coarsely granulated. They belong to mononuclear phagocyte septum. They are abundant in renal lymphomyeloid tissue and spleen in Oncorhynchus mykiss. The tissue bound macrophages are termed as reticuloendothelial septum (RES), which is a system of primitive cells, from which monocytes originate.

Macrophages are present in various other tissues of fishes such as pronephros and olfactory mucosa, etc. Macrophages system of the spleen, bone marrow and liver play a role in phagocytosis of RBC which undergo degradation. Irons separated from haemoglobin molecule are remove by the liver.

Granulocytes

These cells possess specific granules in large numbers and they retain their nucleus.

They are of three types:

    a. Neutrophils

    b. Eosinophils

    c. Basophils

a. Neutrophils:

The neutrophils in fishes are most numerous of the white blood cells and constitute 5-9% of total leucocyte in Solvelinus fontinalis. They are 25% of total leucocytes in brown trout. They are named for their characteristic cytoplasmic staining.

They can be easily identified by the multi-lobed shape of their nucleus and therefore they are segmented or multi-lobed but in some fishes neutrophils are bilobed.

Microphotograph showing neutrophils in
the blood of notopterus notopterus X 1000.

Their cytoplasmic granules are pink, red or violet in peripheral blood smear or azurophilic. Peroxide is present in the azurophil granules of neutrophils and is responsible for bacterial killing. Neutrophils possess Golgi apparatus, mitochondria, ribosomes, endoplasmic reticulum, vacuoles and glycogen. In Carassius auratus, three types of granules are reported in the neutrophils.

The nucleus often looks like the human kidney. In Giemsa stained smear the nucleus is reddish violet in colour and usually exhibits a reticular structure with heavy violet colour. Neutrophils show peroxidase and sudan black positive reaction. The neutrophil is an active phagocyte. It reaches to inflammation site and inflammation refers to local tissue responses to injury.

b. Eosinophils:

There is much controversy regarding the presence or absence of eosinophils, but both eosinophils and basophils function as antigen sensitivity stress phenomenon and phagocytosis. They occur in low percentage.

These cells are generally round and cytoplasm contains a granule which has affinity to acidic dye and they take deep pinkish orange or orange red with purple orange background. The nucleus is lobed, takes deep orange purple or reddish purple stain.

c. Basophils:

The basophils are round or oval in outline. The cytoplasmic granules take deep bluish black stain. They are absent in anguilled and plaice.

Thrombocytes or Spindle Cells

These are round, oval or spindle shaped cells, hence called thrombocytes but in mammals they are disc-like and are called platelets.

They occupy as much as half of the total leucocytes in fish. It constitutes 82.2% of WBC in herring but only 0.7% in other teleosts. The cytoplasm is granular and deeply basophilic in centre and pale and homogenous on the periphery. The cytoplasm takes pinkish or purplish colour. The thrombocytes help in clotting of blood.

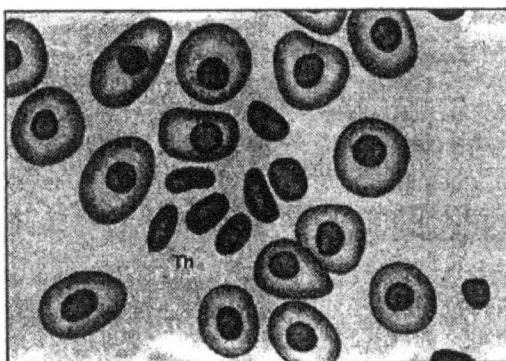

Microphotograph showing thrombocytes
in the blood of notopterus X 1000.

Formation of Blood Cells (Haemopoiesis) in Fishes

The formation of cells and fluid of the blood is known as haemopoiesis, but usually the term hae-mopoiesis is restricted to cells. In mammals the first phase of haemopoiesis or haematopoiesis in developing individual occurs in "Blood Island" in the wall of yolk sac. It is followed as hepatic phase, i.e., the haemopoietic centres are located in the liver and in lymph tissues.

The third phase of foetal haemopoiesis involves bone marrow and other lymphatic tissues. After birth haemopoiesis occurs in red bone marrow of the bones and other lymphatic tissue. It is generally agreed that fishes lack haemopoietic bone marrow. Hence erythrocytes and white blood corpuscles are produced in different tissues.

There are two theories, the monophylatic theory and dualistic or polyphyletic theory regarding the origin of vertebrate blood. According to polyphyletic theory, the blood cells arise from a common stem cells whereas according to the monophylatic theory each blood cell arises from its own blood cells. Recent experiments indicate that monophylatic theory is correct or widely accepted.

The vertebrate blood arises from pleuropotent stem cells in haemopoietic tissue. Such experiments are lacking in fishes. Piscine stem cells founded on analogy with mammalian haemopoiesis are in-direct morphological evidence. Both RBCs and WBCs are originated from lymphoid haemo-blast or haemocytoblast usually mature after they enter the blood stream.

In fishes apart from spleen and lymph nodes many more organs take part in the manufacture of the blood cells. In elasmobranch fishes, the erythrocyte and granulopoietic tissues are produced in the organ of Leydigs, the epigonal organs and occasionally in the kidney.

The Leydig organ is whitish tissue and analogous to bone marrow-like tissue found in oesophagus but the main site is spleen. If the spleen is removed, then the organ of Leydig takes over erythro-cyte production.

In teleost, both erythrocytes and granulocytes are produced in kidney (pronephros) and spleen. The teleost spleen is distinguished into a red outer cortex and white inner pulp, the medulla. The erythrocytes and thrombocytes are made from cortical zone and lymphocytes and some granulo-cytes originate from medullary region.

In higher bony fishes (actinopterygii) red blood cells are also destroyed in the spleen. It is not known whether other organs also function into blood decomposition or how blood destruction comes about in the jawless fishes (Agnatha) or in the basking shark and rays (Elasmobranchii).

In Chondrichthyes and lungfishes (Dipnoi), the spiral valve of the intestine produces several white blood cell types. RBC and WBC are formed from haemocytoblast precursor cells which originate from a variety of organs but usually mature after they enter the blood stream (or immature blood corpuscles are usually of two types, large and small).

Function of Blood Cells

The blood of fishes like other vertebrates consists of cellular components suspended in plasma. It is a connective tissue and is complex non-Newtonian fluid. The blood is circulated throughout the body by cardiovascular system. It is circulated chiefly due to the contraction of heart muscles. The blood performs several functions.

A few important functions are mentioned as follows:

a. Respiration: An essential function is the transport of dissolved oxygen from water and from the gills (respiratory modifications) to the tissue and carbon dioxide from the tissue to the gills.

b. Nutritive: It carries nutrient material, glucose, amino acids and fatty acids, vitamin, electrolytes and trace elements from alimentary canal to the tissue.

c. Excretory: It carries waste materials, the products of metabolism such as urea, uric acid, creatine, etc. away from cell. Trim ethylamine oxide (TMAO) is present in all fishes. It is in high concentration in marine elasmobranchs. Creatine is an amino acid which is the end product of metabolism of glycine, arginine, and methionine whereas creatine is formed by spontaneous cyclisation of creatine. Its level in plasma is 10-80 μm and is excreted by kidney.

d. Haemostasis of Water and Electrolyte Concentration: The exchange of electrolyte and other molecules and their turn over is the function of blood. Blood glucose levels often cited is being a sensitive physiological indicator of stress in fish and there is no unanimity about the blood glucose levels amongst fishes.

e. Hormones and Humoral Agent: It contains regulatory agent such as hormones and also contains cellular or humoral agent (antibodies). The concentration of various substances in the blood is regulated through feedback loops that sense changes in concentration and triggers the synthesis of hormones and enzymes which initiate the synthesis of substances needed in various organs.

Digestion and Absorption

Digestion is the process by which ingested materials are reduced to molecules of small enough size or other appropriate characteristics for absorption, i.e., passage through the gut wall into the blood stream. This generally means that proteins are hydrolyzed to amino acids or to polypeptide chains of a few amino acids, digestible carbohydrates to simple sugars, and lipids to fatty acids and glycerol. Materials not absorbed are by definition indigestible and are eventually voided as faeces. Digestibility ranges from 100 percent for glucose to as little as 5 percent for raw starch or 5-15 percent for plant material containing mostly cellulose (plant fibre). Digestibility of most natural proteins and lipids ranges from 80 to 90 percent.

Digestion is a progressive process, beginning in the stomach and possibly not ending until food leaves the rectum as faeces. Most studies of digestion simply compare the protein, lipid and carbohydrate content of the faeces with that of the feed. A study on digestion in channel catfish showed continuing digestion (and absorption) of protein during passage through each part of the gut. The comparison of faeces collected from the rectum and from the water also points out the hazard of incomplete recovery of faecal matter being likely when collection is done from outside the gut. Most of the protein digestion occurred in the stomach, but also continued in the intestine.

Table: Apparent Digestibility of Protein by Channel Catfish.

Feed	Stomach	Upper intestine	Lower intestine	Rectum	Trough
20% protein	61.6	65.4	75.0	80.9	96.7
40% protein	61.4	72.2	86.5	96.6	98.3

Temperature and pH play major roles in determining the effectiveness of digestive enzymes as a whole. Although most enzyme production decreases at temperatures above or below acclimation temperature, most enzyme activity (for a given amount of enzyme) increases in proportion to the temperature over a wide range of temperatures.

In general, enzyme reaction rates continue to increase at higher temperatures, even though the temperatures increase beyond the lethal temperatures for the species, until the enzymes begin to denature around 50-60 °C. On the other hand, enzymes have limited ranges of pH over which they function, often as little as 2 pH units. Data for channel catfish are probably representative of many teleosts. Acid concentrations (pH) in the stomach ranged from 2 to 4, then became alkaline (pH = 7-9) immediately below the pylorus, decreased slightly to a maximum of 8.6 in the upper intestine, and finally neared neutrality in the hindgut. Fish having no stomach have no acid phase in digestion.

The site of secretion in teleost stomachs appears to be a single kind of cell which produces both HCl and enzyme(s). This contrasts with mammals where two types of cells occur, one for acid and one for enzymes. The production of acid in teleosts is presumably the same as in mammals - NaCl and H_2CO_3 react to produce $NaHCO_3$ and HCl, with the blood being the source of both input materials, which are later mostly reabsorbed in the intestine. One possible explanation for the loss of stomachs in some species of fish is that they live in a chloride-poor environment and that providing large amounts of chloride ion for operating a stomach is bioenergetically disadvantageous. In addition to acid and enzymes, the stomach wall also secretes mucus to protect the stomach from being digested. As long as the rate of mucus production exceeds the rate at which it is washed and digested away, the gut wall is protected from being digested. When mucus production slows or fails, e.g. during gut stasis, during stressful conditions, or post mortem, the gut wall can be eroded or even perforated by the gut's own digestive enzymes.

Two sites produce enzymes in the midgut - the pancreas and the intestinal wall. The intestinal wall is folded or ridged in simple patterns which can be species specific. Secretory cells for both mucus and all three classes of enzymes develop in the depths of the folds, migrate to tops of the ridges (closest to the gut lumen), and then discharge their products. The pancreatic cells produce enzymes and an alkaline solution which are delivered to the upper midgut through the common bile duct. The control of pancreatic secretions (and the pyloric sphincter) in fish is probably the same as in mammals, but there is no information on teleosts yet.

The physical state of food passing through the gut varies with species and type of food. Fish, such as salmonids, which eat relatively large prey, reduce the prey in size layer by layer. Gastric digestion proceeds in a layer of mucus, acid, and enzyme wherever the stomach wall contacts the food. Food appears liquified only in the midgut and solidifies somewhat again during formation of faeces. Pellets of commercial feed seem to be treated similarly, i.e., pellets get smaller and smaller in size with time, although stomachs of some recently-fed salmonids have been found to contain moderate amounts of liquified pellets. Stomachs of juvenile Pacific salmon captured in the open sea contained a thick slurry of pieces of amphi-pods in various stages of solubilization. Fish whose food contains high levels of indigestible ballast, e.g., common carp feeding on a mixture of mud and plants, probably show minimal change in the appearance or volume of their food while it passes through the gut. Microphagous fish, such as the milkfish (Chanos) whose food starts out as a suspension of fine particles, probably also keep it in much the same form all the way through the gut. In general: there seems not to be the same degree of liquifaction of food in fish as is commonly described for mammals.

Absorption of soluble food could begin in the stomach - it occurs in mammals, but has not been investigated in fish - but takes places predominantly in the midgut and probably to some degree in the hindgut. The sites and mechanisms of absorption are largely unstudied, except histologically. Several histologists have identified fat droplets in intestinal epithelial cells following a lipid-rich meal. Increased numbers of leucocytes in general circulations following a meal by the sea bream and increased number of fat droplets in them have been described. It was hypothesized that leucocytes entered the gut lumen, absorbed lipid droplets, and then returned to the blood stream. It is clear that the mammalian type of villi with their lymph duct (lacteal) inside are absent in fish, although there is some folding and ridging of the gut wall to increase surface area. Lacteals serve as a primary uptake route in mammals for uptake of droplets of emulsified lipids (chylomicra). Teleost fish have a lymphatic system which includes extensions into the gut wall, but its role in lipid uptake is unknown. Absorption of amino acids, peptides, and simple carbohydrates has been little studied, but presumably they diffuse through or are transported across the gut epithelium into the blood stream. What light microscopists identified as a brush border on the surface of the epithelial cells facing the gut lumen, has now been clarified with electron-microscopy as microvilli; i.e., subcellular, finger like projections of the cell membrane who's greatly increased surface area is probably involved in absorption.

Digestive System in Fishes

The digestive system consists of alimentary canal and its associated glands. The digestive tube also contains numerous intramural glands which provide the tube by lubricating mucus, enzymes, water, etc. The extramural glands are liver, pancreas and gall bladder.

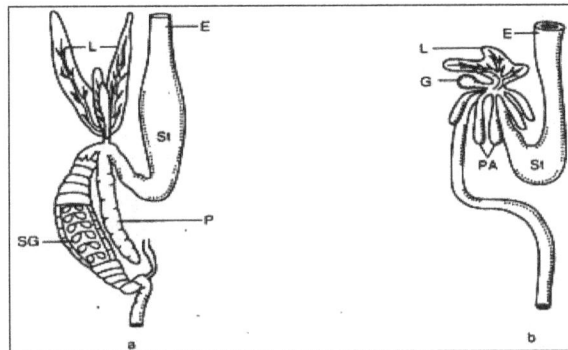

Diagram of alimentary canal of (a) Elasmobranch,(b) teleost.
E, oesophagus; G, gallbladder, L, liver; P, pancreas; PA, phyloric
appendices; SG, spiral gut; St stomach.

The liver is present in all fishes. The pancreas which is exocrine and endocrine organ may be a discrete organ or it may be diffused in the liver or in the alimentary canal. In sharks and rays (Elasmobranchii) pancreas is relatively compact and usually well developed as a separate organ, often two lobed, but in teleosts, the pancreas is diffused in the liver to form hepatopancreas.

It is also diffused in the alimentary canal in a few fishes. It is also present in the mesenteric membranes surrounding the intestine and liver. The gall bladder is vestigeal in deep sea fishes but it is prominent in other fishes.

While passing through the alimentary canal, the food is broken down physically and chemically and ultimately solubilized so that degraded products can be absorbed. The absorption occurs chiefly through the wall of intestine.

The undigested food and other substances within the alimentary canal such as mucus, bacteria, desquamated cells and bile pigments and detritus are excreted as faeces. The peristaltic movement and local contractions are important and help the food to pass through the gut. The local contraction displaces intestinal contents proximally and distally.

Parts of Alimentary Canal

The alimentary canal of fish consists of mouth, which opens into buccopharynx, which in its turn opens into the oesophagus. The oesophagus opens into the stomach/intestine. The lips, buccal cavity and pharynx are considered as non-tubular part whereas the oesophagus stomach/intestinal bulb, intestine and rectum are tubular in nature and distinguished as tubular part of the alimentary canal.

Mechanism of Feeding

In most teleost, the food reaches the mouth by sucking it by enlarging their buccal and opercular cavities. The pressure in the buccal and opercular cavities and pressure of water around the fish, are important for sucking. The pressure of − 50 to -105 cm of water was recorded, as food was sucked in, and + 1 to + 9 cm of water taken in with the food was driven out through operculum.

In case of black bullhead (Ictalurus) a negative pressure of − 80 cm was recorded at 18 °C. It is also recorded that to produce strong negative pressure, all muscles principally involved have to exert tension near maximum isometric tension. Osse made an electromyography study of feeding in Perca and confirmed the sequence of muscle action inferred from anatomical studies and by analogy with respiration.

Stimuli for Feeding

The mechanism of feeding behaviour in fishes is very complicated. There are generally several kinds of stimuli for feeding. The common factors affecting the internal motivation or drive for feeding include season, time of the day, light intensity, time and nature of last feeding, temperature and any internal rhythm.

The visual, chemical, taste and lateral line system also control the momentary feeding act. The interaction of these groups of factors determines when and how a fish will feed and what it will feed upon.

The role of visual and olfactory factors in connection with feeding behaviour has been studied by experimental conditions by Groot. He found visual, chemical and mechanical sense organs in Pleuronectidae, Soleidae, and Bothidae (belongs to the family of flat fishes, Pleuronectiformes).

Soleidae are polychaet mollusc feeder, feed during night, find their food mainly by olfactory clues, but still have the capability of finding their food visually. The barbels help the fish to locate food grubbed from soft bottom material.

Table: Role of visual and olfactory factors in feeding.

Family	Type	Way of finding food	Form of intestine	Form of gill raker	Olfacto-ry Lobe	Optic Lobes
Bothidae	Fish-feeder	Vision loop	Simple toothed	Heavily	Small	Large
Pleuro nectidae	Crustaceafeeder	Mainly vision but also olfaction	Complicated loop	Less toothed	Medium	Large
Soleidae	Polychaetamolluscafeeder	Olfaction but vision possible	More complicated loop	Few or no teeth	Large	Small

On the basis of feeding habits, the fishes are categorized as follows:

1. Herbivorous

2. Carnivorous

3. Omnivorous

4. Detrivorous

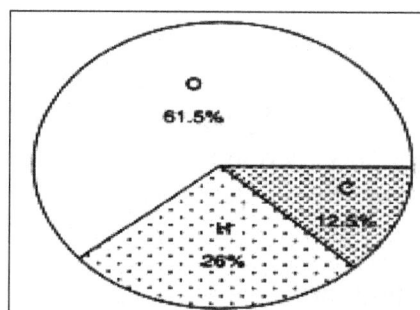

Diagram showing distribution of species with different feeding habits . Carnivorous; H, herbivorous. C. omnivorous.

The fishes can be classified as 'Euryphagous' consuming mixed diet 'Stenophagous—eating limited assorted food and 'Monophagous'—consuming only one type of food. Amongst teleosts, about 61.5% are omnivorous, 12.5% are carnivorous and about 26% are herbivorous.

Herbivorous Fishes

They consume about 70% unicellular algae, filamentous algae and aquatic plants. In addition to plant material these fishes also consume 1-10% of animal food and mud. The common examples are Labeo species Osphronemus goramy, Sarotherodon mossambicus etc. Herbivorous fishes have long and coiled intestine.

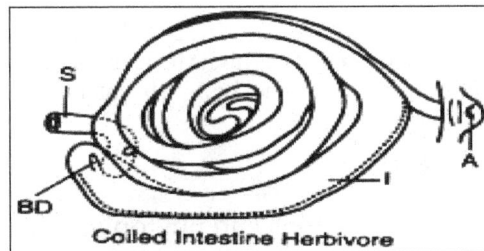

Diagram of alimentary canal of a herbivorous catfish.
A anus; BD, bile duct; I, intestine; S, stomach.

Carnivorous Fishes

The fishes in contrast to herbivore have shorter gut, the intestine is straight, very little coils are present. Some of the carnivores possess intestinal caecae. They prey on small organisms and consume high percentage of animals such as copepods, dafnia and insects.

The examples of carnivorous fishes are Walla go attu, Mystus seenghala, Mystus cavassius, Mystus vittatus, Channa striatus, Channa marulius, Channa punctatus, Notopterus chitala, Rita rita, etc.

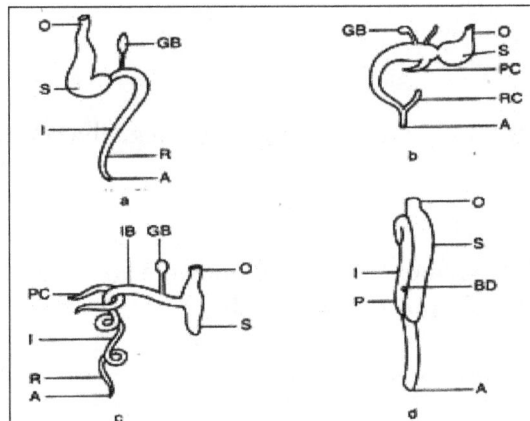

Alimentary canal of fishes. (a) Mystus seenghala; (b) Notopterus chitala; (c)Channa striatus
(d) A carnivore fish, Esox luaus. A, anus; BD, bile duct; CB, gall bladder; I, intestine; IS, intestinal bulb;
o, oesophagus; P, pylorus; PC, pyloric caecum; P. rectum; RC, rectal caecum; S, stomach.

Omnivorous Fishes

Omivorous fishes like Cyprinuscarpio, Cirrhina mrigala, Puntius, Clarias, etc. are consuming both plants and animals. The rotifers, mud and sand are also found in the alimentary canal. Their gut length intermediates between carnivorous and herbivorous fishes.

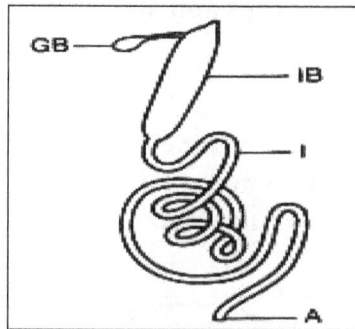

Alimentary canal of cirrhinus mrigala. A, Anus;
I. intestine IB, intestinal bulb; GB, gall bladder.

Detrivorous or Plankton Feeder

They consume detritus along with zooplanktons and phytoplankton's. The arrangement of gill rakers is such that it filters them from water. The examples are Catla catla, Hilsa ilisa, Cirrhina reba, and Hypopthalmichthys molitrix. They are both omnivorous and carnivorous.

The fishes can also be named on the basis of modification of buccopharynx

1. Predator
2. Grazers
3. Strainers
4. Suckers
5. Parasitic

Predators

They possess well developed grasping and holding teeth, e.g. sharks, pike and gars, etc.

Grazers

They take the food by bite. These fishes feed on plankton and on bottom organisms, e.g., bluegill (Lepomis macrochirus), parrot fish and butterfly fish.

Strainers

They have efficient straining or filtering adaptation due to the arrangement of gill rakers forming sieve for straining the food material. They are plankton feeders.

Suckers

The fishes have inferior mouth and sucking lips. The response depends upon stimulus of touch e.g., sturgeons, Labeo, Osteochilus , etc.

Parasites

Amongst fishes, the deep sea eel (Simenchelys parasiticus) is parasitic in nature. Lamprey and hagfish are parasitic but belongs to cyclostomata.

Reproduction

Reproduction is the process by which new individual organisms are produced. Reproduction is a fundamental feature of all known life; each individual organism exists as the result of reproduction. Although the term reproduction encompasses a great variety of means by which organisms produce new offspring, reproductive processes can be classified into two main types: Sexual reproduction and asexual reproduction.

The methods of reproduction in fishes are varied, but most fishes lay a large number of small eggs, fertilized and scattered outside of the body. The eggs of pelagic fishes usually remain suspended in the open water. Many shore and freshwater fishes lay eggs on the bottom or among plants. Some have adhesive eggs. The mortality of the young and especially of the eggs is very high, and often only a few individuals grow to maturity out of hundreds, thousands, and in some cases millions of eggs laid.

Males produce sperm, usually as a milky white substance called milt, in two (sometimes one) testes within the body cavity. In bony fishes a sperm duct leads from each testis to a urogenital opening behind the vent or anus. In sharks and rays and in cyclostomes the duct leads to a cloaca. Sometimes the pelvic fins are modified to help transmit the milt to the eggs at the female's vent or on the substrate where the female has placed them. Sometimes accessory organs are used to fertilize females internally-for example, the claspers of many sharks and rays.

In the females the eggs are formed in two ovaries (sometimes only one) and pass from the ovaries to the urogenital opening and to the outside. In some fishes the eggs are fertilized internally but shed before development takes place. Members of about a dozen families each of bony fishes (teleosts) and sharks bear live young. Many skates and rays also bear live young. In some bony fishes the eggs simply develop within the female, the young emerging when the eggs hatch (ovoviviparous).

Others develop within the ovary and are nourished by ovarian tissues after hatching (viviparous). There are also other methods utilized by fishes to nourish young within the female. In all live-bearers the young are born at a relatively large size and are few in number. In one family of primarily marine fishes, the surfperches from the Pacific coast of North America, Japan, and Korea, the males of at least one species appear to be born sexually mature, although they are not fully grown.

Some fishes are hermaphroditic, an individual producing both sperm and eggs, usually at different stages of its life. Self-fertilization, however, is probably rare. Successful reproduction and in many cases defense of the eggs and young is assured by rather stereotyped but often elaborate courtship and parental behaviour, either by the male or the female or both. Some fishes prepare nests by hollowing out depressions in the sand bottom (cichlids, for example), build nests with plant materials and sticky threads excreted by the kidneys (sticklebacks), or blow a cluster of mucus-covered bubbles at the water surface (gouramis). The eggs are laid in these structures. Some varieties of cichlids and catfishes incubate eggs in their mouths.

Some fishes, such as salmon, undergo long migrations from the ocean and up large rivers to spawn in gravel beds where they themselves hatched (anadromous fishes). Others undertake shorter migrations from lakes into streams or in other ways enter for spawning habitats that they do not ordinarily occupy.

Types of Reproduction

Generally two types of reproduction are seen, these are:

- Sexual reproduction
- Asexual reproduction

Sexual Reproduction

A form of reproduction that involves the fusion of two reproductive cells (ova and sperm) in the process of fertilization normally especially in animals, it requires two parents, one male and the other female. The female of the species produces an egg which is fertilized by sperm from the male. Depending on the species, fertilization takes place either within the female's body or externally, after the female lays her eggs and the male releases his sperm close by. When the egg is fertilized the genes of both parents are combined to produce zygote which develops into a new individual. Nearly all animals reproduce sexually.

- Bisexual: Which is the prevalent kind; sperm and eggs develop in separate male and female individuals.

- Hermaphrodites: Both sexes are in one individual and, as among certain errands and a dozen or more other families. Self-fertilization or true functional hermaphrodites' sexists. Hermaphroditic sex glands are known for several species, including some trout relatives (salmonoids), perches (Perca), walleyes (Stizostedion). Darters (Etheostoma). And some of the black basses (Micropterus).

Sexual reproduction is a process of biological reproduction by which organisms give rise to descendants that have a combination of genetic material contributed by two different gametes, usually i.e. two different individuals, male and female. A gamete is a mature reproductive or sex cell. Sexual reproduction results in increasing genetic diversity, since the union of these gametes produces an organism that is not genetically identical to the parent(s).

Sexual reproduction is characterized by two processes: meiosis, involving the halving of the number of chromosomes to produce gametes; and fertilization, involving the fusion of two gametes and the restoration of the original number of chromosomes. During meiosis, the chromosomes of each pair usually cross over to achieve genetic recombination. Once fertilization takes place, the organism can grow by mitosis.

While typically sexual reproduction is thought of in terms of two different organisms contributing gametes, it also includes self-fertilization, whereby one organism may have "male" and "female" parts, and produce different gametes that fuse.

Sexual reproduction is the primary method of reproduction for the vast majority of visible organisms, including almost all animals and plants. The origin of sex and the prominence of sexual reproduction are major puzzles in modern biology.

Gametes, Meiosis, Fertilization and Mitosis

In the first stage of sexual reproduction, 'meiosis', a special type of that takes place in gonads prior to the production of gametes, the number of chromosomes is reduced from a diploid number (2n) to a haploid number (n). During 'fertilization', haploid gametes come together to form a diploid zygote and the original number of chromosomes (2n) is restored in the offspring.

Sexual reproduction involves the fusion or fertilization of gametes from two different sources or organisms.

Typically, a gamete or reproductive cell is haploid, while the somatic or body cell of the organism is diploid. A diploid cell has a paired set of chromosomes. Haploid means that the cell has a single set of unpaired chromosomes, or one half the numbers of chromosomes of a somatic cell. In diploid organisms, sexual reproduction involves alternating haploid (n) and diploid (2n) phases, with fusion of haploid cells to produce a diploid organism. Some organisms, however, exhibit polyploidy, whereby there are more than two homologous sets of chromosomes.

Meiosis and mitosis are two diverse types of cell divisions. Mitosis occurs in somatic (body) cells. The resultant number of cells in mitosis is twice the number of original cells. The number of chromosomes in the daughter cells is the same as that of the parent cell. Meiosis occurs in reproductive or sex cells and results in gametes. It results in daughter cells with half the number of chromosomes present in their parent cell. Essentially, a diploid cell duplicates its chromosomes, then undergoes two divisions (tetroid to diploid to haploid), in the process forming four haploid cells. This process occurs in two phases, meiosis I and meiosis II.

Fertilization involves the fusion of haploid gametes to give a diploid organism, which can then grow by mitosis. Thus, in sexual reproduction, each of two parent organisms contributes half of the offspring's genetic makeup by creating haploid gametes that fuse to form a diploid organism. For most organisms, a gamete that is produced may have one of two different forms. In these anisogamous species, the two sexes are referred to as male, producing sperm or microspores as gametes, and female, producing ova or megaspores as gametes. In isogamous species, the gametes are similar or identical in form, but may have separable properties and may be given other names. For example, in the green alga, Chlamydomonas reinhardtii, there are so-called "plus" and "minus" gametes. A few types of organisms, such as ciliates, have more than two kinds of gametes.

Sexually reproducing organisms have two sets of genes (called alleles) for every trait. Offspring inherit one allele for each trait from each parent, thereby ensuring that offspring have a combination of the parents' genes. Having two copies of every gene, only one of which is expressed, allows deleterious alleles to be masked.

As with the other vertebrates, sexual reproduction is the overwhelming dominant form of reproduction. However, there are several genera of fish that practice true or incomplete parthenogenesis, where the embryo develops without fertilization by a male.

Although vertebrates in general have distinct male and female types, there are fish species that are both males and females gonads (hermaphrodites), either at the same time or sequentially. For example, the amenone fish spend the first part of their lives as males and later become females, and the parrot fish is first female and then male. Some members of the Serranidae (sea basses) are simultaneous hermaphrodites, such as the Serranus and their immediate relatives, Hypoplectrus (the synchronoous hermaphroditic hamlets).

Fertilization may be external or internal. In the yellow perch, eggs are produced by ovaries in the female and sperm is produced by the testes, and they are released through an opening into the environment, and fertilization takes place in the water. In some live bearers, such as guppies and swordtails, females receive sperm during mating and fertilization is internal.

Other behaviors related to sexual reproduction include some species, such as the stickleback, built nests from plants, sticks, and shells, and many species that migrate to spawn.

Asexual Reproduction

Reproduction in which new individuals are produced from a single parent without the formation of gametes. Asexual reproduction enables animals to reproduce without a partner. It occurs chiefly in lower animals, microorganisms, and plants. Many invertebrates reproduce asexually, including coral and starfish. Coral grow small buds that break off to become separate organisms, whilst starfish are able to generate an entirely new being from a fragment of their original body.

- Other Types: Without above classification, another three types of reproduction are also possible.

- Parthenogenesis: Is the development of young without fertilization, and a condition that has been called parthenogenesis (more properly gynogenesis) occurs in a tropical fish, the live-bearing Amazon molly (poecilia Formosa), and is also known in poeciliopsis. Mating with a male is required, but the sperm serves only one of its two functions, that of inciting the eggs to develop; it does not take any part in heredity. The resultant young are always females, with no trace of parental characters.

Asexual reproduction is a form of reproduction in which an organism gives rise to a genetically-similar or identical copy of itself without a contribution of genetic material from another individual. It does not involve meiosis, ploidy reduction, or fertilization, and only one parent is involved genetically. A more stringent definition is agamogenesis, which refers to reproduction without the fusion of gametes.

Asexual reproduction is the primary form of reproduction for single-celled organisms such the archaea, bacteria, and protists. However, while all prokaryotes reproduce asexually (without the formation and fusion of gametes), there also exist mechanisms for lateral gene transfer, such as conjugation, transformation, and transduction, whereby genetic material is exchanged between organisms. Biological processes involving lateral gene transfer sometimes are likened to sexual reproduction. The reproductive variances in bacteria and protists also may be symbolized by + and - signs (rather than being called male and female), and referred to as "mating strains" or "reproductive types" or similar appellations.

Many plants and fungi reproduce asexually as well, and asexual reproduction has been cited in some animals, including bdelloid rotifers, which only are known to reproduce asexually, and various animals that exhibit parthenogenesis under certain conditions. In parthenogenesis, such as found in some invertebrates and vertebrates, an embryo is produced without fertilization by a male. Generally, parthenogenesis is considered a form of asexual reproduction because it does not involve fusion of gametes of opposite sexes, nor any exchange of genetic material from two different sources however, some authorities classify parthenogenesis as sexual reproduction on the basis that it involves gametes or does not produce an offspring genetically identical to the parent (such as a female domestic turkey producing male offspring).

A wide spectrum of mechanisms may be exhibited. For example, many plants alternate between sexual and asexual reproduction and the freshwater crustacean Daphnia reproduces by parthenogenesis in the spring to rapidly populate ponds, then switches to sexual reproduction as the intensity of competition and predation increases. Many protists and fungi alternate between sexual and asexual reproduction.

A lack of sexual reproduction is relatively rare among multicellular organisms, which exhibit the characteristics of being male or female. Biological explanations for this phenomenon are not completely settled. Current hypotheses suggest that, while asexual reproduction may have short term benefits when rapid population growth is important or in stable environments, sexual reproduction offers a net advantage by allowing more rapid generation of genetic diversity, allowing adaptation to changing environments.

Hermaphroditism

A hermaphrodite is defined as any individual organism that possesses both male and female reproductive organs during their life span. The main advantage of hermaphroditism is the assurance of a reproductive partner. Although hermaphroditism is quite common in invertebrates and plants, it is an exceedingly rare occurrence in vertebrates. Hermaphroditism in mammals and birds are almost always a pathological condition (often leading to infertility). Only in Perciforms (fish) does hermaphroditism occur naturally and in high frequency.

Hermaphrodites are divided into two main categories: synchronous hermaphrodites, and sequential hermaphrodites. In the synchronous hermaphrodites, organisms possess both active male and active female reproductive organs at the same time. In sequential hermaphrodites, both male and female reproductive organs may be present, but only one is active and viable at any given time.

Synchronous or simultaneous hermaphrodites in reef fish are relatively atypical. A few Serranids (sea basses, e.g. Serranus sp.) and Hamlets are known synchronous hermaphrodites. During mating, one individual will lay eggs while another fertilizes the eggs, after which both will reverse roles and perform fertilization again. Synchronous hermaphrodites do not fertilize themselves; Self-fertilization does not promote genetic diversity, and can amplify genetic defects from parent to offspring. The interesting fact is most synchronous hermaphrodites form monogamous pairs.

Sequential hermaphrodites are so named because they are capable of transforming from one sex to another. Theses transformations entail a full conversion of gonads from one sex to another. The gonads of sequential hermaphrodites possess the genetic information to produce both male and female reproductive organs, but only the dominant gene is expressed at any giventime. Different cues - varying from species to species - may induce sex changes.

Sequential hermaphrodites are further categorized into two main categories: protogynous and protandrous. Protandrous hermaphrodites are those that develop into males first, then possibly to females. Protogynous hermaphrodites are the exact opposite, with juveniles first developing female reproductive organs that may possibly change into male reproductive organs in select circumstances. It should be noted that hermaphrodites do not necessarily have to change sexes, but by definition, are capable of this feat.

Protandrous hermaphrodites are the rarer of the two types. Pomacentrids (damselfish) are the most famous of these hermaphrodites. For example, clownfish of the genus Amphiprion live in communities that consist of one dominant female specimen and several smaller male (or asexual juvenile) specimens. If the female should be removed, a male will convert to a female, insuring a reproductive partner for the community.

Protogynous hermaphrodites are most often haremic fish. These fish form monoandric harems comprising of 1 male overseeing numerous females for life. The two primary responsibilities of the male are to defend its territory against other conspecific males, and to court and fertilize females of its territory. If the male should die (either of natural causes or conflict-related mortality), the dominant female of the harem will undergo a sex change from female to male. This sex change may take as little as 5 days. The new male will then resume the full responsibilities of the previous male until he should die. Protogynous hermaphrodites that form harems include the wrasses of the genus Cirrhilabrus and Paracheinlinus, Dwarf Angelfish of the genus Centropyge, and Anthias (e.g. Pseudoanthias sp.).

Some protogynous fish do not form harems, but may form pairs. The dottybacks (Pseudochromis sp.) are presumably protogynous hermaphrodites that fall under this category.

There are also sequential hermaphrodites that waver between sexes with no discernable order. The sexes of these fish are often determined by the ratio of sexes in an immediate community. These types of hermaphrodites include numerous gobies.

Modes of Reproduction and Development

The ability to produce new living individual is a basic characteristics of all plants and animals. All reliable evidence indicates that new life comes only from pre-existing life; this is the process of biogenesis or reproduction. The different modes of reproduction can be classified according to the environment in which the embryos develop, and the sources of the nutrients supporting embryonic growth. These are mentioned into the following term:

Oviparity

Oviparity or egg laying refers to the situation where the development or the fertilized of egg occurs outside the body of the female. The young hatch when the egg envelope, shell or capsule is broken. Oviparous fish may be further categorized as being either ovuliparous or zygoparous.

Ovuliparity refers to the release of ova from the reproductive tract of the female followed by fertilization or activation in the external environment. Thus, all organisms that have external fertilization and this includes most teleost are said to be ovuliparous.

Zygoparity refer to the oviparous condition in which the zygotes (i.e. fertilized ova products of fusion between the eggs and sperm) are retained within the body of the female for a short period of time before being released into the environment. Obviously, zygoparous species display internal fertilization with their being a transfer of male sperm to the reproductive tract of the female. Zygoparous reproduction characterizes all skates some sharks and a small number of teleost.

Irrespective of whether the fertilization of the eggs occurs internally or the egg yolk provides externally the nutrient for the developing embryos of oviparous species.

Oviparity in Fishes

Amongst fish which reproduce by ovipartity, there are several groups.

Egg Scatterers

Fish which breed in this way either spawn in pairs or in groups. Males and females release milt (the sperm and spermatic fluid of the male) and eggs into the water at the same time. These mix together, fertilising the eggs. The fertilised eggs are broadcast (or spread) into the plankton column and float away in the current or sink to the bottom. No parental care is given, so large amounts of eggs are produced. It is easy to produce many eggs and because they are in the water, they don't dry out - necessary oxygen and nutrients aren't scarce. When theoffspring settle out of the plankton, they might be in totally new environments: this gives the young a chance to survive across a wide area. The main disadvantage of this method is that the fish must go through a larval stage before they transform into adults. In this larval stage, they are very vulnerable while they try to find food and avoid predators. Also, they may not find a suitable environment when they settle out of the plankton column. The survival rate for individual eggs is very low, so the parent has to produce millions of eggs.

Egg Depositors

These fish either lay eggs on a flat surface, like a stone or plant leaf or may even place them individually among fine leaved plants. The parents usually form pairs and guard the eggs and fry (young fish) from all danger. The Cichlids such as Koi are the best known species for this. Some Catfish and Rainbow fish are also egg depositors.

Nest Builders

Many fish species build nests. These might be a simple pit dug into gravel (trout do this) or an elaborate bubble nest. When they are ready to spawn, the fish may construct a nest by blowing bubbles, and they often use vegetation to anchor the nest. The male will keep the nest intact and keep a close eye on the eggs. The Gouramis, Anabantids and some catfish are the most common of this type.

Mouthbreeders

These are particularly odd, since eggs are fertilised externally, but raised internally. The females usually lay their eggs on a flat surface where they are then fertilised by the male. After fertilization the female picks up the eggs and incubates them in her mouth. Broods tend to be small, since by the time the fry are released by their mothers they are well formed and suffer minimal losses. The best known mouthbreeders are the African lake Cichlids.

- Relative fecundity: is the number of eggs produced in a season per unit somatic weight of the fish (i.e., eggs/gram), and is useful if it is shown that the fecundity of a fish is proportional to its weight, which is not uncommon.

- Population fecundity: is the number of eggs spawned by the population in one season, is the sum of the fecundities of all females, and is usually expressed as the product of the expected fecundity of an average female, the number of breeding females in the population (an example of why classifying maturity stages may be useful).

Spawning

Spawning refers to the release of unfertilized planktonic eggs by female fish, which is the reproductive pattern for most marine fishes. The eggs are fertilized shortly after release by males. Some fishes also deposit unfertilized Fish Reproduction eggs in nests where they are fertilized and develop. Fishes with internal fertilization release free-swimming larvae, or juveniles. The ripening of eggs and spawning are controlled by hormones, nutrition of the female and external (ecological) factors. Usually maturation and spawning are controlled by a combination of endogenous and exogenous controls and are not governed by any specific factor.

Terminology used in discussing spawning:

- Mating: pairing (one-on-one) for the purpose of fertilizing eggs; copulatory organ present.

- Spawning: release of unfertilized eggs into the environment or release of larvae into the environment; mating and spawning need not occur simultaneously (e.g., surfperches). Spawning can occur without true mating (e.g., herring, which are broadcast spawners).

- Fertilization: fusion of eggs and sperm (creating diploids from haploids); mating and fertilization need not occur simultaneously (e.g., surfperches and rockfishes).

- Incubation time: time from egg fertilization to hatching. Gestation applies only to live-bearing fishes; it is the time young stay within the female.

- Hatching: when the larva frees itself from the egg.

- Breed: to produce offspring by hatching (or gestation).

- Brood: guard and groom eggs until they hatch.

Maturity Stages

Male

- Stage I (Immature): The testes are extremely small, often recognized only as translucent filamentous strands in the early stages. Later, when easily visible, they are small, opaque, pinky-white, leaf-like structures with a fairly long vas deferens, which get easily snapped when the testes are removed. There is very little asymmetrical development and, if any, the left testis is slightly longer. The testes with the vas deferens occupy roughly 50 % of the length of the body cavity or occasionally slightly more, measuring a maximum overall length of about 35 mm, of which the testicular portion may form 10 to 15 mm only. The gonad weight ranges up to 0.2 gm, but it is usually less than that. Their relative weight to body weight is normally below 0.8%.

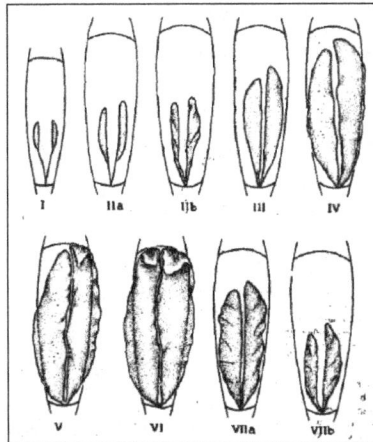

Diagrammatic representation of different maturity stages of male.

- Stage II-A (Developing Virgin): The testes are thicker and more elongated. They are opaque, pink or white in colour, with vas deferens reduced but thread-like. Asymmetrical development has set in and the left testis is almost always longer. The testes with the ducts extend to 50-60% of the body cavity and measure 30 to 40 mm and the testicular portion measures about 25 to 30 mm. Their weight ranges from 0.2 to 0.5gm, but it is usually around 0.4gm. The relative weight varies from 1 .0 to 1.7%.

- Stage II-B (Spent-resting): The testes are pinkish or brownish-white in colour but different from the previous stage in having shrunken and wrinkled appearance when viewed against light, showing that the organs are not compactly filled with germ cells. The left testis is usually longer. The vas deferens is a much wider duct than the thread-like passage of the previous stage and is covered by the lower halves of the testes which are narrower and extend almost to the posterior end of the body cavity. With the degree of opacity varying, there are a few patches of semi-opaque regions. The organs fill up 50 to 55% of the body cavity and measure almost the same length as the previous stage, i.e., 30 to 40 mm, but theif absolute and relative weight are much less. While the former ranges up to 0.29 gm with an average of about 0.15 gm, the latter is about 0.5%.

- Stage III (Maturing): The testes are well developed and thickened, white in colour. Vas deferentia, being filled with spermatogonia, are reduced and measure less than 15 mm. The left testis is distinctly longer and this condition persists in the subsequent stages also. The gonads varying in length from 36 to 50 mm occupy 70 to 75 % of the body cavity. Their weight ranges up to 2.0 gm but it is usually around 1.0 gm forming 2.5 to 4-0% as relative weight.

- Stage IV (Maturing) : Quite massive and creamy-white are the testes with vas deferens hidden under them. The organs, while extending to the entire breadth of the body cavity, occupy 85 to 90% of the body cavity length, measuring 40 to 70 mm. Their absolute weight ranges from 2.5 to 5.0 gm, but in majority of cases it is around 3.5 gm. Their relative weight varies from 5.0 to 7.0%.

- Stage V (Mature): The testes are opaque white in colour, soft, more extensive than Stage IV and occupy the entire length of the body cavity but very often even more with the result their anterior ends tend to fold down along the ventral body wall. On a little pressure internally at the posterior end, spermatic fluid oozes out. Their usual length is 65 to 70 mm but can attain even 80 mm sometimes. Their absolute weight ranges from 6.0 to 8.0 gm, with an average around 7.0 gm. The relative weight works up to 10.0 to 14.0%.

- Stage VI (Running): The testes are very extensive, white in colour and fill the entire space of body cavity displacing the intestines to a fraction of space. Not only the anterior tips are folded down along the ventral body wall but even the outer margins extending along the sides of the body wall curve towards the middle so much so very often an insertion along the mid-ventral line cuts through the outer edges of the testes. Under a slight pressure externally on the flanks of the fish or even while handling, milt extrudes out. The organs always measure more than 70 mm and weigh from 9.0 to 13.0 gm with an average around 10.0 gm and the relative weight ranges between 15 and 20%.

- Stage VII A (Partially Spent): The testes are meat-coloured, a bit leathery in texture, shrunken with wrinkles and semi-opaque spaces visible when viewed against fight. Measuring 40 to 60 mm, they occupy 70 to 80% of the body cavity and weigh usually 2.0 gm but may range from 1.0 to 2.5 gm, which forms 2.5 to 4.5% as relative weight.

- Stage VII B (Spent): The testes are deep flesh-coloured, shrunken, flat, strap-like, and shrivelled with translucent patchy regions. They occupy 50 to 60% of the body cavity and measure 30 to 45 mm. Their weight is around 0.5 gm, forming about 1.5% as relative weight.

Female

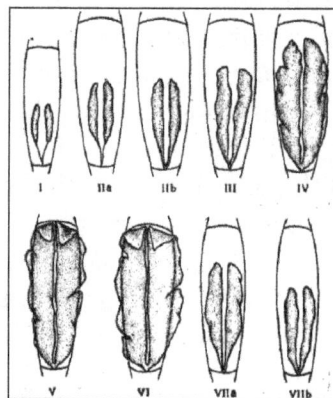

Diagrammatic representation of different maturity stages of female.

- Stage I (Immature): The ovaries are soft cylindrical and almost translucent, pink or flesh-coloured. Sometimes due to post-mortem changes they appear purple in colour. The surface of the ovary is smooth with no distinct blood vessels. There is very little asymmetry in the size of the ovaries. The oviduct is fairly long and completely transparent with the result, the ovarian bodies look like detached stubs, short and plump. The entire length of the ovaries with their ducts occupies about or slightly more than 50% of the body cavity and measures up to about 35 mm of which the length of the ovary alone ranges from 10 to 25 mm. Their absolute weight may be about or below 0.25gm forming a maximum of 0.8% as relative weight. The ovaries are compactly filled with oocytes, not visible to naked eye. The oocytes are yolkless and transparent and measure up to 0.13 mm with the majority of them in the size range of 0.07 to 0.09 mm.

- Stage II A (Developing Virgin): Cylindrical, soft, translucent ovaries pink or flesh-coloured. Asymmetry is not quite distinct yet. Oviducts, thin and thread-like, are a little reduced in length and not more than 10 mm. The overall length of the gonads ranges from 30 to 35 mm forming 55 to 60% of body cavity. The ovaries alone measure about 20 to 25 mm in length and usually weigh around 0.4 gm, but may range from 0.3 to 0.6 gm. The relative weight is 1.0 to 1.5%. Majority of the ova are transparent with signs of yolk formation in some, which are mostly semi opaque but sometimes fully opaque with or without translucent periphery. However, even these do not appear as distinct grains to be easily recognised with naked eye. Maximum diameter of ova recorded is 0.30 mm with a large number of ova ranging in size from 0.15 to 0.18 mm.

- Stage II B (Spent-resting): The ovaries are dark-red or brownish red or deep flesh-coloured, having a collapsed and flattened appearance. External surface is wrinkled. The tunica is thicker and the Oviducts much wider and shorter than in the previous stage. The length of the organs is 30 to 45 mm occupying 55 to 60% of the body cavity. Their weight is commonly around 0.2 gm, forming less than 0.5% as relative weight. Occasionally, late spawners resting in January-February period may record a maximum gonad weight of about 0.4 to 0.5 gm with their relative weight of about 0.8%. Majority of ova are transparent, not visible to naked eye and measure 0.07 to 0.11 mm. A few scattered opaque ova may be present without transparent periphery and measure up to 0.15 mm. This stage is characteristically distinguished by the presence of clots of blood cells appearing as brownish masses in between the oocytes.

- Stage III (Maturing): The ovaries are turgid, opaque and yellow in colour with granular appearance. Development of blood vessels is perceptible. The oviducts are very much reduced. Usually there is asymmetrical development in the size of the ovaries, either gonad longer than the other. This condition persists in the subsequent stages also. The ovaries occupy about 65 to 70% of the body cavity and measure from 35 to 50 mm. Their weight ranges from 0.8 to 1.5 gm, but usually it is below 1.0gm. The relative weight may accordingly vary from 2 to 4%. The maximum diameter of ova recorded is 0.54 mm. The size frequency distribution of ova normally shows one distinct mode of maturing ova around 0.35 to 0.40 mm, which are opaque with translucent periphery and visible to naked eye. In these ovaries, some semi-opaque ova with yolk deposition around the centre measure 0.15 to 0.17 mm and form a good percentage. Sometimes, ovaries advanced a little further but not entered into Stage IV show two modes of maturing ova, an advance one 0.42 to 0.47 mm and a minor one around 0.22 to 0.27 mm. Both sets of ova are opaque and are provided with translucent periphery.

- Stage IV (Maturing): The ovaries are compact, vascular with conspicuous blood vessels on the tunica and bright yellow in colour. Oviducts are not quite distinct. The organs almost extend to the entire body cavity forming 80 to 90% of the latter's length, with its own length varying from 45 to 65 mm. They weigh about 2.0 to 4.5 gm with an average of 3.0 gm, forming 4.5 to 7.0% as relative weight. The largest ova may measure about 0.67 mm and the size frequency polygons show two distinct modes, one Anywhere between 0.51 and 0.57 mm and the other around 0.27 to 0.34 mm. The former are completely opaque, while the latter are provided with translucent periphery.

- Stage V (Mature): The ovaries are orange-yellow in colour, fully vascular with prominent blood vessels ramifying on the surface. Tunica is very thin and bursts at slight pressure.

Ovaries are very often more than the length of the body cavity with the anterior tips taking a loop down. Their normal length is 65 to 70 mm, but according to the size of fish may even extend up to 80 mm. Their average weight is 6.5 gm with a range of 5.5 to 7.5 gm. Relative weight ranges from 9 to 13%. Maximum diameter of ova observed is 0.82 mm. The distribution of ova shows two groups, an advanced mature one anywhere between 0.62 and 0.67 mm. and a maturing group around 0.35 to 0.47 mm. The ova of former group Present varying appearances; some are completely opaque, some provided with narrow or wide transparent periphery, some vacuolated, some partly transparent and a few fully. Completely transparent ova have one big Oil globule or two-three smaller droplets of oil globule which measure from 0.054 to 0.109 mm and the other partly transparent and vacuolated ova have number droplets of oil globule. The ova of less advanced mode are fully opaque.

- Stage VI (Running): The ovaries appear as a sort of cream-coloured cellophane bags filled with boiled sago. At a slight prick, a gelatinous mass of transparent ova flows out, the tunica being so thin. Ova can be extruded on slight pressure externally on the flanks of the fish or even while handling. The ovaries measure more than 70 mm and fill the entire-space of abdomen cavity displacing the intestine to a narrow space in between the two ovaries- They may weigh from 8.0 to 12.0 gm, but ordinarily around 9.0 gm, forming 13 to 17% as relative weight. The largest ova are transparent and jelly-like reaching a maximum diameter of 1.2 mm, but the majority of these range from 0.80 to 0.91 mm in diameter with one or two oil globules very rarely cleaved into three which measure 0.091 to 0.127 mm. There is a significant number of medium-sized ova forming another distinct mode between 0.46 and 0.51 mm, which are completely opaque or provided with transparent periphery.

- Stage VII A (Partially Spent): The ovaries are dark red in colour either throughout the length or only at the posterior half they are a bit flaccid and collapsed with slight wrinkles on the surface. The ovarian lamellae are clearly seen as book leaves especially at the posterior region indicating that the lamellae are not compactly filled with maturing ova and that some have been shed. The ruptured blood capillaries produce blood clots which appear as brownish or reddish masses. Blood capillaries penetrate deeply into the interior. Although the ovaries are shrunk in volume and reduced in breadth, they extend to 70 to 80% of the body cavity measuring 40 to 60 mm. Their weight ranges from 1.5 to 3.0 gm but is usually around 2.0gm, amounting to 2.5 to 5.0% as relative weight. The maximum diameter of ova is about 0.60 mm and the frequency distribution shows only one distinct mode anywhere within 0.35 to 0.51 mm. These ova are completely opaque but a few of these perhaps in the process of resorption are translucent with greyish yolk in the centre within which may be found a few droplets of oil-globule.

- Stage VII B (Spent): The ovaries are elongated, honey-coloured, bloodshot, flabby, limp and gelatinous with wrinkles on surface due to collapsed condition. The tunica is leathery and the wide oviduct is now Recognizable. The ovaries measure about 30 to 45 mm and occupy 55 to 60% of the body cavity. They almost always weigh around 0-5 gm with a maximum limit of 1.0gm, forming 0.6 to 1.5% as relative weight. Recently, spent fish have remnants of mature ova, measuring a maximum diameter of 0.47 mm, as resorbing and disintegrating opaque structures. The blood cells from ruptured capillaries appear as reddish clots. Sometimes, there may be a few scattered droplets of oil globule. The resorbing eggs are sometimes translucent in appearance, with the yolk in the form of small spherules, light grey or brownish in colour with man droplets of oil globule clustering around the centre. These ova form a small mode in the size frequency distribution around 0.27 to 0.31 mm. At a later stage, a few blood-coloured or brownish masses are seen which represent the unspawned ova broken down and covered up by the blood cells. In this state, the ovaries appear deep flesh-coloured. The rest are all immature transparent oocytes less than 0.25 mm in diameter.

Factors Triggering Maturation and Spawning

There are three primary factors that influence the events leading up to spawning: nutritional state of the female, physiological factors (hormones), and ecological factors.

Nutrition of the Female

The feeding condition of the mother can have an important effect on the final maturation of the eggs. Two examples from Hempel show that in some of the Atlantic herring populations spawning may occur only every other year if environmental conditions, particularly those affecting food supply, are poor. Also, it has been found in the laboratory that in Atlantic sole (Solea solea) no spawning occurred when the flatfish were fed a diet (mussels only) deficient in certain amino acids; however, when the flatfish were force-fed the missing amino acids they spawned, indicating the ovary had been unable to obtain the needed amino acids from maternal tissue when the nutrition of the female had been inadequate.

Physiological Factors

Hormones govern migration and timing of reproduction, morphological changes, mobilization of energy reserves, and elicit intricate courtship behavior. The pituitary is the major endocrine gland that produces gonadotropin, which controls gametogenesis, the production of gametes, namely sperm (spermatogenesis) and eggs (oogenesis), by the gonads. The pituitary also controls the production of steroids (steroidogenesis) by the gonads; once the gonads are stimulated by the pituitary they begin producing steroids, which in turn control yolk formation (vitellogenesis) and spawning. The control of spawning by the pituitary is often used in fish farming such as in the production of caviar from sturgeon (Acipenser spp.) where spawning is induced by injecting pituitary extract at a late stage of gonadal development, usually in combination with changes in temperature and light periodicity.

Ecological Factors

Often ecological factors are associated with timing so that food availability is optimal for the larvae. Some ecological factors important to spawning are temperature, photoperiod, tides, latitude, water depth, substrate type, salinity, and exposure.

- Temperature: An important factor in determining geographical distributions of fishes. Although little is known about the mechanism by which temperature controls maturation and spawning in fishes, for many marine and freshwater fishes the temperature range in which spawning occurs is rather narrow, so that in higher latitudes the minimum and maximum temperature requirement for spawning is often the limiting factor for geographical distribution and for the successful introduction of a species into a new habitat. For example,

Pacific halibut (Hippoglossus stenolepis) are found spawning primarily in areas with a 3-8 °C temperature on the bottom and therefore do not spawn in Puget Sound, although the adults are caught in the northern areas of Puget Sound. In fact, even in highly migratory tuna, spawning is restricted to water of specific temperature ranges.

- Photoperiod and periodicity: The day length (photoperiod), in some cases at least, is thought to influence the thyroid gland and through this the fishes' migratory activity, which is related to gonadal development (maturation). In the northern anchovy, by combining the effects of temperature and day length, continued production of eggs under laboratory conditions was brought about by keeping the fish under constant temperature conditions of 15 °C and a light periodicity of less than 5 hours of light per day (Lasker personal communication). In high latitudes, spawning is usually associated with a definite photoperiod (and temperature), which dictates seasonal pulses of primary production in temperate regions to assure survival of larvae. In low latitudes, where there is little variation in day length, temperature, and food production, other factors may be important such as timing with the monsoons, competition for spawning sites, living space, or food selection.

Reproductive periodicity among fishes varies from having a short annual reproductive period to being almost continuous. There is a tendency for the length of the reproductive period to shorten with increasing latitude. Thus tropical fishes spawn nearly continuously, whereas subarctic fishes spawn predictably during the same few weeks each year. Presumably times of spawning have evolved so larval development will coincide with an abundant food supply. Within spawning seasons, fish may spawn on a daily or monthly tidal cycle or on a diel cycle, or in association with some other environmental cue, such as a change in daylength, temperature, or runoff. A notable instance of spawning periodicity associated with the tidal cycle is the California grunion (Leuresthes tenuis), which spawns intertidally at the peak of the spring high tides. Within species, spawning times may vary with latitude: Generally, in species that spawn as daylength increases, spawning occurs earlier in the year in lower latitudes than at higher latitudes. In species that spawn as daylength decreases, spawning takes place earlier in the year at higher latitudes than at lower latitudes.

- Tides (moon cycles): The dependence of spawning on moon cycles in California grunion spawning on California beaches is an extreme example of external factors controlling reproduction in fishes. Grunions are adapted to spawning on the beach every two weeks in the spring during a new or full moon. Spawning is just after the highest high tide; therefore, eggs deposited in the sand are not disturbed by the surf for 10 days to a month later. Eggs will hatch when placed in agitated water (which simulates surf conditions). In Puget Sound, surf smelt (Hypomesus pretiosus) spawn year-round, except in March. Surf smelt deposit eggs at high tide in sand and gravel (but not necessarily at the highest tide). On the open ocean shores, spawning occurs at midtidal heights (for different subpopulations) (Dan Pentilla personal communication).

- Latitude and locality: Pacific herring show a definite relationship between latitude and spawning time. Spawning is early in San Francisco (December, January); later in Washington State (February, March, April, May); and still later in Alaska (April, May, June). These fishes are perhaps of different, distinct subpopulations. In temperate waters a bimodal distribution of eggs is usually seen, which indicates discontinuous spawners. The smaller-sized mode represents resting eggs for a future spawning, and the larger mode represents maturing eggs (oocytes), which will presumably be spawned within the year. Temperate water fishes are also usually deterministic, which means all eggs to be spawned are determined at the start of the year. A polymodal distribution of eggs is typical of tropical areas and some temperate water fishes, which signifies continuous or serial spawners, and indicates several spawnings. A well-known temperate example would be Pacific sardine (Sardinops sagax), which spend 7 months spawning and 2 months developing/maturing. Batch spawning has been described for northern anchovies Tropical spawners are usually nondeterministic, which means the eggs to be spawned are not determined at the start of the year but are produced throughout the year; however, nondeterministic can also represent the spawning potential for successive years, an example being the Atlantic cod (Gadus morhua), which will have several years' spawn in the ovary.

In general, older fish usually spawn first and younger fish later, which means that a prolonged spawning period for a population may not be true for individual fish. Once a set of eggs is mature and hydrated, the female may release them all at once or in several batches. An example of releasing several batches is plaice (Pleuronectes platessa), where a single female two weeks after releasing one batch of eggs releases more eggs, and then three weeks later she releases the remaining eggs. Another example is the Pacific herring, which spawn once a year and females lay about 100 eggs per spawning act, which they repeat several hundred times over a few days. In the lab, walleye pollock spawned an average of nine times in an average period of 27 days.

It also needs mentioning that a long duration of the spawning season of a population cannot necessarily be taken as an indication of prolonged spawning of the individual fish. The prolonged period may be due to differences in spawning time between age groups since older fish tend to spawn earlier in the season. Furthermore, the coexistence of different spawning subpopulations must be taken into account, since winter and summer spawners may be distinct stocks, although shifts from one seasonal spawning pattern to the other may occur. An example of how unpredictable this can be is that certain Atlantic herring of low fecundity have been found to always spawn in the winter, regardless of whether they originated from winter or summer spawning.

- Water depth: Pacific herring spawn along beaches, marine grasses, and algae. Atlantic herring do not spawn along shore but in deeper water up to 200 m (the clearest difference between the Pacific and Atlantic herring, which are usually, designated distinct species on the basis of genetic analysis). Of course fishes often spawn at one depth but live at different depths during other times of the year. For example, petrale sole (Eopsetta jordani), in which spawning occurs in a specific offshore area 300-400 m deep, were found by fishermen and eventually had to be protected with regulations to prevent overfishing (A.C. Delacy personal communication).

- Spawning substrate type: Pacific herring spawn on vegetation whereas Atlantic herring spawn on solid substrate (e.g., gravel). Lingcod spawn on rocks, pilings, and cracks in solid substrate; this species protects the egg mass. Some species such as buffalo sculpin (Enophrys bison) and plainfin midshipman (Porichthys notatus) spawn intertidally and will stay with the egg mass even when they are exposed at a low tide.

- Salinity: Also a factor affecting spawning. There are varying salinities in many areas of estuaries. Some species will shift spawning sites because of salinity changes. Various degrees of mixing, precipitation, and freshwater runoff may alter spawning habits.

- Exposure and temperature: A clear example of shifting spawning sites in response to temperature and exposure is the black prickleback (Xiphister atropurpureus), where spawning is shifted from winter in protected areas to spring in exposed areas. The complex effects of lower or higher wave action and lower or higher temperatures on courtship, gonadal development, and spawning behavior that result in the spawning site shift.

Growth

Growth is a characteristic feature of living beings. Every organism grows and attains its normal size. Growth is actually an increase in any dimension of any organism in relation to time. Growth is the process of addition of flesh as a result of protein synthesis.

Knowledge of fish growth is of vital importance for obtaining high yield of fish. The rate of growth varies from species to species and sometimes it varies even among species also.

Rate of Fish Growth is influenced by Many Factors

1. Different Localities: Individuals of same species grow at different rate in different localities.

2. Seasonal Effects: The different temperatures in different seasons affect growth of fish. In summer the fish grow fast, however, in winter growth is slow.

3. Availability of Food and Oxygen: This is the most important factor influencing growth of fish. In ecologically balanced water body, growth rate is high because of ample supply of oxygen and food.

4. Population Density: Higher population density shows slow growth while low density shows high growth rate.

5. Age: Fish continue to grow throughout their life but with increasing age, growth rate is slow and in extreme old age fish growth is extreme slow. Growth slows down at the onset of sexual maturity when large amount of nutrient material periodically go into egg or sperm formation.

6. Adaptive Character: Sharks and Sturgeon grow very fast into large size because of constant food supply, which provides protection against their predators. However, in Goby growth is slow, resulting in small size of the body. This is an adaptation to overcome problem of limited food supply and is related to increased reproductive capacities as they have more danger of destruction by predators.

Growth Curve

Growth curve is obtained when either length or weight of a fish is plotted against time periods. The curve obtained is sigmoid or S-shaped. This curve illustrates the rate of growth. If rate of growth is plotted against time, such curve is known as growth rate curve.

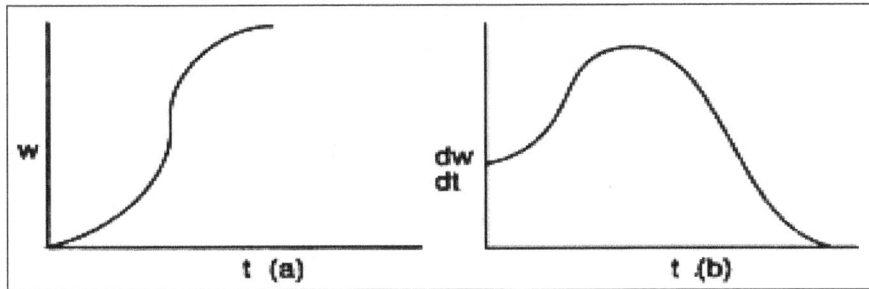

(a) Growth curve, (b) Growth rate curve.

Linear Growth

Fish growth varies according to various periods. Maximum growth rate is found just before attaining the sexual maturity of fish. After maturation of gonads the growth slows and decreases. A linear growth pattern is seen during the postnatal development of fish, especially during the period of hatching to the sexual maturity.

This may differ from species to species, and depends on various factors, viz., appropriate and balanced food, optimum temperature for vital activities like metabolism, vitamins and trace elements for proper growth. Fluctuation in these factors affects growth rate of the fish. If there is scarcity of food the size of fish will be small, and, if the availability of food is restored to normalcy, the fish grows to normal.

Determination of Length (Linear Growth)

The scales of many fishes have permanent marks, which is considered as confirmation of both age and growth. This evidence provides the base of estimating yearly changes in growth and also of calculating the length of body. Annuli formed during winter shows zone of retarded growth.

There exists a relationship between the annuli and the growth of the fish. Thus body length can be estimated by studying the relationship between the scale size and the body size.

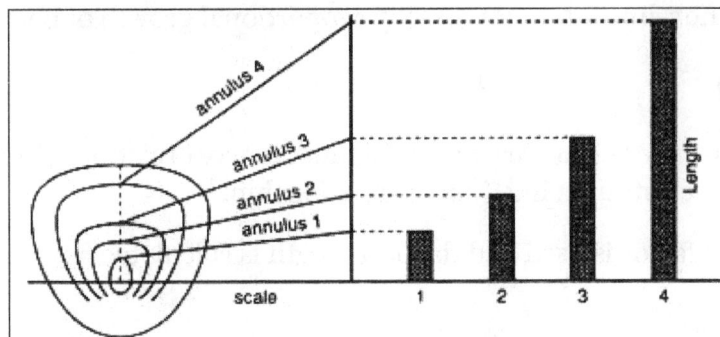

Back calculation of length.

Some important methods are as follows:

Direct Proportion—Body/Scale Relationship Method

This method is based on the facts of isogonic growth rate, which shows that the growth of the scale and the body occurs at the same proportional rate. However, the calculated body length is found to be less than its empirical length. This is because of the fact that fish attains some length before scale formation.

Following formula is applied to calculate the length of a fish:

$$\frac{S^+}{S} = \frac{L^+}{L}$$

Where,

S^+ = Length of scale radius to annulus

S = Length of total radius scale

or,

$$L^+ = \frac{S^+}{S} \times L$$

Where,

L^+ = Length of fish when annulus X was formed

L = Length of fish when scale sample was obtained

This method is useful in calculating the length of fish at early stages. The above formula works well when fish shows a linear and directly proportional body. For example if a fish has L = 600 mm, S=10 mm, S^+ = 4 mm, then the length of the fish at the time of formation of first annulus can be calculated as follows:

$$L^+ = \frac{S^+}{S} \times L = \frac{4 \times 600}{10} = 240\,mm$$

Here the fish shows a non-linear but not directly proportional growth of body and scale.

Dahl-Lea Method

This method states that there is a curvilinear (sigmoid curve) relationship between body length and scale length and not a straight line (linear) relationship.

Therefore, a correction factor is used and the past length is calculated as:

L = C S,

Where,

L = body length, S = scale length and C = a constant,

or,

log L = log C + n log S,

where,

C and n = constants from data.

It is observed that body/scale relationship varies from species to species.

Length-weight Relationship

Weight can be calculated by a formula:

$$W = KL^3,$$

Where,

W = weight, L = length, K = constant; for fish showing symmetrical or isometric growth throughout.

The body form and specific gravity of a fish change constantly with the advancement of age. Thus simple cube W expression therefore does not seem to be correct throughout the life history of fish, as the value of K is not constant but subject of great variation.

A more satisfactory formula is:

$$W = L^n,$$

Where,

W = weight, L = length, n = constants for log W = log C + n log L.

Linear Growth and Condition of Fish: Coefficient of Condition Factor

The ratio of length and growth represents the condition at any given time. When this ratio is calculated and if the obtained value is large, it indicates better condition of fish. The Fulton's formula is used to express condition of the fish.

$$Q = W100,$$

Where,

Q is coefficient of condition.

I^3 W is the weight of fish and I is the length.

Coefficient factor or coefficient condition is that state in which a fish remains for a certain period.

This value is useful in elucidating the following:

- Difference in conditions of the individuals of same species in different water.

- Differences among individuals of same length.

- Differences occur due to seasonal changes in relation to the age and sex of the fish.

There is another way to know the condition of the fish by estimating the fat contents, which is related to linear growth of fish. During the period of winters, migration and spawning, the major fat contents are saturated fatty acids. This type of fat is used for long- term use. On the other hand, unsaturated fatty acids are present in fat when the fat is deposited for use in daily maintenance.

Food Consumption and Growth of Fish

Food plays an important role in the growth of fish.

Food consumption rate is estimated by the following methods:

1. Samples of fish are collected from the nature. Then quantity of food present in the stomach is determined followed by estimation of rate of gastric digestion.

2. By determining the intake of nitrogen, which helps in the growth, and the loss of nitrogen in excretion through kidney and gills.

3. By comparison of food consumption and growth rate between the fish from natural habitat and the fish kept in captivity. It has been found that the total amount of food consumed is not used for growth. Only a small amount of food is utilized for growth.

Hence it is necessary to know about the fate of consumed food, which is as follows:

1. Some part of food is neither digested nor absorbed and is passed out along with faeces.

2. A part of ingested food is digested and absorbed by the gut.

3. Some part of absorbed food is excreted through the kidney, gall bladder and skin.

4. Some part of assimilated food is used for maintenance of metabolism.

5. Some part of assimilated food is used for growth.

Life Span of Fishes

It has been observed that generally species with large individuals seems to exhibit older age than species consists of small fish. Many fishes that attain size of 12 inches show a life span of at least 4 to 5 years. However, age of fishes in captivity, i.e., in the aquarium may be expected to be more but do not exceed to 5 years.

Factors affecting Fish Growth

Abiotic Factors

Fishes show a considerable variation in their growth rate and age of sexual maturity in northern and southern waters. Temperature plays an important role as it affects metabolism and food consumption in fishes. The seasonal increase in the temperature of freshwater bodies cause increase in natural food for fishes.

Fishes grow at faster rate at the optimum temperature, at which their appetite is high and requirement of maintenance is low. The optimum temperature varies from 7 °C to 19 °C in case of brown trout (Solmo trutta).

Light also affects the rate and pattern of fish growth. Intensity of light depends on the length of the day. Light causes increased thyroid activity resulting the increased swimming activity and reduction in growth rate in experiments with the brown trout (Salmo trutta) in mid-summer. Longer days are advantageous where feeding is accomplished by sight whereas shorter days are beneficial when the food is abundant.

Biotic Factors

The population density is also of a vital importance. Among the fish population there is competition for food, nest site and so occur, which may affect the growth of fish. High density of fish in a place can also affect the fish growth. It has been described that the fish in a crowded condition may release a chemical substance known as "crowding substance", which prevents growth.

Genetic Factor

Growth and size of the fish are genetically determined. Size difference between the sexes may be due to genetic factors.

Endocrine Factor

Several metabolic hormones like insulin, thyroxin, gonadal and adrenal steroids produce synergy which affects the growth. In carefully controlled experiments, Higgs, treated groups of yearling coho salmon with bovine growth hormone thyroxin and methyltestosteron, given separately and with the combination of these three substances maximum growth was obtained.

Immune Response

The immune system can be defined as a complex system that protects the organism against organisms or substances that might cause infection or disease. One of the most fascinating characteristics of the immune system is its capability to recognize and respond to pathogens with significant specificity. Innate and adaptive immune responses are able to recognize foreign structures and trigger different molecular and cellular mechanisms for antigen elimination. The immune response is critical to all individuals; therefore numerous changes have taken place during evolution to generate variability and specialization, although the immune system has conserved some important features over millions of years of evolution that are common for all species. The emergence of new taxonomic categories coincided with the diversification of the immune response. Most notably, the emergence of vertebrates coincided with the development of a novel type of immune response. Apparently, vertebrates inherited innate immunity from their invertebrate ancestors.

Fish are a heterogeneous group divided into three classes: Agnatha (jawless fish such as the hagfish and lampreys), Chrondrichthyes (cartilaginous fish such as sharks, rays and skates) and Osteichthyes (bony fish). As in all vertebrates, fish have cellular and humoral immune responses and organs, the main function of which is immune defence. Most generative and secondary lymphoid organs in mammals are also found in fish, except for lymphatic nodules and bone marrow.

Immune responses in fish have not been as well characterized as they have in higher vertebrates. Consequently, there is not enough information about the components of the fish immune system and its function and regulation. Key immune mammalian homologous genes have been identified in several fish species, suggesting that the fish immune system shares many features with the mammalian system. For example, the identification of α and β T cell receptor genes (TCR), key T cell markers such as CD3, CD4, CD8, CD28, CD40L, and a great number of cytokines and chemokines suggest that T helper $(Th)_1$, Th2 and Th_{17} and the regulatory subset Treg are present in fish. Some cell subsets have been better studied mainly because their activity can be easily differentiated and measured, as in the case of cytotoxic cells and macrophages. Finally, B cells have been much more studied due to the availability of monoclonal antibodies that have been isolated and identified by a number of techniques. Phenotypic characterization of leukocytes has been hampered mainly by the lack of membrane cell markers. Researchers anticipate developing antibodies for cell lineage markers of fish immunocompetent cells that can be used to isolate and characterize immune cells to obtain insights into their regulation and role in immune response.

Antibodies in teleosts play a key role in the immune response. In general, IgM is the main immunoglobulin in teleosts that can elicit effective specific humoral responses against various antigens. For IgM, one gene alone can generate as many as six structural isoforms. Therefore, diversity is the result of structural organization rather than genetic variability. Recently, several reports have provided evidence for the existence of IgD/IgZ/IgT in fish. Interestingly, B cells from rainbow trout and salmon have high phagocytic capacity, suggesting a transition in B lymphocyte during evolution in which a key cell type of the innate immunity and phagocytosis evolved into a highly specialized component of the adaptive immune response in higher vertebrates.

Innate and Adaptive Immune Response

The development of an immune system is essential for the survival of living organisms. In vertebrates, immunity can be divided into two components, the innate immune response and the adaptive immune response. The innate immune response is the initial line of defence against infection, which includes physical barriers and cellular response. The adaptive immune response is capable of specific antigen recognition and is responsible for the secondary immune response.

The innate immune system recognizes conserved molecular structures common to pathogenic microorganisms such as polysaccharides, lipopolysaccharides (LPS), peptidoglycans, bacterial DNA, and double-strand viral RNA, among others, through their interaction with specific receptors like toll receptors (TLRs). These mechanisms of recognition may lead directly to successful removal of pathogens, for instance by phagocytosis, or may trigger additional protective responses through induction of adaptive immune responses. Cells of the innate immune system have a diverse array of functions. Some cells are phagocytic, allowing them to engulf and degrade pathogenic particles. Other cells produce and secrete cytokines and chemokines that can stimulate and help guide the migration of cells and further direct the immune response.

The adaptive system recognizes foreign structures by means of two cellular receptors, the B cell receptor (BCR) and the T cell receptor (TCR). Adaptive immunity is highly regulated by several mechanisms. It increases with antigen exposure and produces immunological memory, which is the basis of vaccine development and the preventive function of vaccines. The adaptive response generally starts days after infection and is capable of recognizing specific protein motifs of peptides, which leads to a response that increases in both speed and magnitude with each successive exposure. The main effector cells of the adaptive immune response are the lymphocytes, specifically B cells and T cells. When B cells are activated, they are capable of differentiating into plasma cells that can secrete antibodies. Upon activation T cells differentiate into either helper T cells or cytotoxic T cells. Helper T cells are capable of activating other cells of the adaptive immune response such as B cells and macrophages, while cytotoxic T cells upon activiation are able to kill cells that have been infected.

Fish Cytokines

Cytokines are secreted proteins with growth, differentiation, and activation functions that regulate the nature of immune responses. Cytokines are involved in several steps of the immune response, from induction of the innate response to the generation of cytotoxic T cells and the production of antibodies. In higher vertebrates, the combination of cytokines that are secreted in response to an immune stimulation induces the expression of immune-related genes through multiple signalling pathways, which contributes to the initiation of the immune response. Cytokines can modulate immune responses through an autocrine or paracrine manner upon binding to their corresponding receptors.

Cytokines have overlapping and sometimes contradictory pleiotropic functions that make their classification difficult. Cytokines are produced by macrophages, lymphocytes, granulocytes, DCs, mast cells, and epithelial cells, and can be divided into interferons (IFNs), interleukins (ILs), tumor necrosis factors (TNFs), colony stimulating factors, and chemokines. They are secreted by activated immune-related cells upon induction by various pathogens, such as parasitic, bacterial, or viral components. Macrophages can secret IL-1, IL-6, IL-12, TNFα, and chemokines such as IL-8 and MCP-1, all of which are indispensable for macrophage, neutrophil, and lymphocyte recruitment to the infected tissues and their activation as pathogen eliminators. Meanwhile, cytokines released by phagocytes in tissues can also induce acute phase proteins, including mannose-binding lectin (MBL) and C-reactive protein (CRP), and promote migration of DCs.

Fish appear to possess a repertoire of cytokines similar to those of mammals. To date several cytokine homologues and suppressors have been cloned in fish species. Some cytokines described in fish are TNFα, IL-1β, IL-6 or IFN.

Current knowledge of fish cytokines is based on mammal models of the cytokines network and their complex interactions.

Pro-inflammatory Fish Cytokines

Tumour Necrosis Factor α (TNFα)

TNFα (tumour necrosis factor alpha) is a pro-inflammatory cytokine that plays an important role in diverse host responses, including cell proliferation, differentiation, necrosis, apoptosis, and the induction of other cytokines. TNFα can induce either NF-kB mediated survival or apoptosis, depending on the cellular context. TNFα mediates powerful anti-microbial responses, including inducing apoptosis, killing infected cells, inhibiting intracellular pathogen replication, and up-regulating diverse host response genes. Many viruses have evolved strategies to neutralize TNFα by direct binding and inhibition of the ligand or its receptor or modulation of various downstream signalling events.

TNFα has been identified, cloned, and characterized in several bony fish, including Japanese flounder, rainbow trout, gilthead sea bream, carp catfish, tilapia, turbot and goldfish. These studies have revealed the existence of some obvious differences from their mammalian counterpart, such as the presence of multiple isoforms of TNFα in some teleost species the high constitutive expression of this gene in different tissues of healthy fish and its relatively poor up-regulation by immune challenge in vitro and in vivo. However, the most unexpected and interesting difference between fish and mammal TNFα concerns the weak in vitro effects of TNFα on phagocyte activation in goldfish, rainbow trout, turbot and gilthead seabram. This weak in vitro activity of fish TNFα sharply contrasts with the powerful actions exerted by the i.p. injection of recombinant TNFα in gilthead seabream, which includes the recruitment of phagocytes to the injection site, with a concomitant strong increase in their respiratory burst. Apparently endothelial cells are the main target cells of fish TNFα, suggesting that TNFα is mainly involved in the recruitment of leukocytes to the inflammatory foci rather than in their activation. Despite the above, differential expression has been observed in studies with rainbow trout leucocytes, which have shown increased response to different pro-inflammatory stimuli, as human recombinant TNFα, LPS, zimosan and muramyl dipeptide as a peptidoglycan constituent of both gram-positive and gram-negative bacteria. Moreover, it is known that Infectious Pancreatic Necrosis Virus (IPNV)-mediated up-regulation of TNFα regulates both the Bad/ Bid-mediated apoptotic pathway and the RIP1 (receptor-interacting protein-1)/ROS-mediated secondary necrosis pathway.

Interleukin 1 Family

In mammals, the 11 members of the Interleukin-1 family include IL-1α (IL-1F1), IL-1β (IL-1F2), IL-1 receptor antagonist (IL-1ra/IL-1F3), IL-18 (IL-1F4), IL-1F5-10 and IL-33 (IL-1F11). These molecules tend to be either pro-inflammatory or act as antagonists that inhibit the activities of particular family members. Despite these semantic issues, to date only two clear homologues of these molecules have been discovered in fish, IL-1β and IL-18.

Interleukin 1β

IL-1β is one of the earliest expressed pro-inflammatory cytokines and enables organisms to respond promptly to infection by inducing a cascade of reactions leading to inflammation. Many of the effector roles of IL-1β are mediated through the up- or down-regulation of expression of other cytokines and chemokine's. Mammalian IL-1β is produced by a wide variety of cells, but mainly by blood monocytes and tissue macrophages. IL-1β was the first interleukin to be characterized in fish and has since been identified in a number of fish species, such as rainbow trout, carp, sea bass, gilt head seabram, haddock, and tilapia. A second IL-1β gene (IL-1beta2) has been identified in trout.

In mammals pro-IL-1β remains cytosolic and requires cellular proteases to release the mature peptide. It is known that the peptide is cleaved by the IL-1β converting enzyme (ICE). However, the aspartic acid residue for which this enzyme has specificity is not present in all fish genes sequenced to date. Nevertheless, using a combination of multiple alignments and analysis of the N-terminal sequences of known mature peptides, it is possible to predict fish gene cutting sites. In trout, this gives a mature peptide of 166 and 165 aminoacids for IL-1β1 and IL-1β2.

Like its mammalian counterpart, teleost IL-1β has been found to be regulated in response to various stimuli, such as LPS or poly I : C. The biological activity of recombinant IL-1β (rIL-1β) has been studied in several fish species, indicating that fish IL-1β is involved in the regulation of immune relevant genes, lymphocyte activation, and migration of leucocytes, phagocytosis and bactericidal activities.

Interleukin 18

In mammals, IL-18 is mainly produced by activated macrophages. It is an important cytokine with multiple functions in innate and acquired immunity. One of its primary biological properties is to induce interferon gamma (IFNγ) synthesis in Th1 and NK cells in synergy with IL-12. It promotes T and NK cell maturation, activates neutrophils and enhances Fas ligand-mediated cytotoxicity. Like IL-1β, it is synthesized as an inactive precursor of approximately 24 kDa and is stored intracellularly. Activation and secretion of IL-18 is mainly effected through specific cleavage of the precursor after D35 by caspase 1, also termed the IL-1β-converting enzyme (ICE), which is believed to be one of the key processes regulating IL-18 bioactivity. Some other enzymes, including caspase 3 and neutrophil proteinase 3, also cleave the IL-18 precursor to generate active or inactive mature molecules.

IL-18 was discovered in fish by analysis of sequenced fish genomes (fugu) and EST databases (medaka). An alternative splicing form of the IL-18 mRNA was discovered in trout that may have an important role in regulating IL-18 expression and processing in this species. This form shows a lower constitutive expression relative to the full length transcript, but unlike the full length transcript, it increases in response to LPS and polyI:C stimulation in the RTG-2 fibroblast cell line. The expression level of the full length transcript can increase in response to LPS plus IL- 1b in head kidney leucocyte cultures, and by IFNγ in RTS-11 cells.

Other Pro-inflammatory Cytokines

Interleukin 6

A number of other interleukins are considered pro-inflammatory, some of which are released during the cytokine cascade that follows bacterial infection. Of this IL-6 is one of the best known, and is itself a member of the IL-6 family of cytokines that includes IL-11 and IL-31, as well as cytokines such as mammalian CNTF, LIF, OSM, CT-1 and CT-2. Whilst the homology of known fish molecules with many of these IL-6 family members is not conclusive, true homologues appear to be present in at least in the cases of IL-6 and IL-11. IL-6 is produced by a diverse group of cells including T lymphocytes, macrophages, fibroblasts, and neurons, endothelial and glial cells. The pleiotropic effects of IL-6 are mediated by a 2-subunit receptor and include the regulation of diverse immune and neuro-endocrine processes. IL-6 has been implicated in the control of immunoglobulin production, lymphocyte and monocyte differentiation, chemokine secretion and migration of leukocytes to inflammation sites.

IL-6 was first discovered in fugu by analysis of the genome sequence and subsequently in other species as part of EST analysis of immune gene-enriched cDNA libraries. However, little is known about the function and signalling pathways of IL-6 in fish. Interestingly, trout IL-6 expression in macrophages is reported to be induced by LPS, poly I:C and IL-1β in the macrophage cell line RTS-11, as well as in head kidney macrophages. Moreover, IL-6 induces the expression of itself, so it can act in an autocrine and paracrine fashion to increase its expression, with the potential to both amplify and exacerbate the inflammatory response. However, IL-6 can significantly down-regulate the expression of trout TNFα1, TNFα2, and IL-1β, suggesting a potential role of trout IL-6 in limiting host damage during inflammation.

Interleukin 11

In mammals, IL-11 is produced by many cell types throughout the body. Basal and inducible IL-11 mRNA expression can be detected in fibroblasts, epithelial cells, chondrocytes, synoviocytes, keratinocytes, endothelial cells, osteoblasts and certain tumour cells and cell lines. Viral and bacterial infection and cytokine stimulation (IL-1, TNFα and TGF-β1] induce IL-11 expression. IL-11 acts on multiple cell types, including hemotopoietic cells, hepatocytes, adipocytes, intestinal epithelial cells, tumour cells, macrophages, and both osteoblasts and osteoclasts. In the hematopoietic compartment IL-11 supports multilineage and committed progenitors, contributing to myeloid, erythroid, megakaryocyte and lymphoid lineages. IL-11 is also an anti-inflammatory cytokine that inhibits the production of pro-inflammatory cytokines from lipopolysaccharde (LPS)-stimulated macrophages. In combination with its trophic effects on the gastrointestinal epithelium, IL-11 plays an important role in the protection and restoration of gastrointestinal mucosa.

The teleostean IL-11 orthologue has been found to consist of duplicate IL-11 genes, named IL-11a and IL-11b, with expression patterns indicating that both divergent forms of teleostean IL-11 play roles in antibacterial and antiviral defence mechanisms of fish. In trout, IL-11 molecule is grouped with IL-11a and is constitutively expressed in intestine and gills and is highly up-regulated at other immune sites (spleen, head kidney, liver) following bacterial infection. In vitro, the macrophage-like RTS-11 cell line has shown enhanced IL-11 expression in response to LPS, bacteria, poly I:C and rIL-1β. In carp, IL-11a is modulated by LPS, ConA and peptidoglycan in head kidney macrophages and cortisol has been found to inhibit IL-11 expression on its own and in combination with LPS. In contrast to carp IL-11a, which shows low levels of constitutive expression in blood leucocytes, IL-11b in Japanese flounder shows higher expression at this site, and strong up-regulation was found in response to rhabdovirus infection in kidney cells. This suggests that these paralogues have some complementarity of function related to their differential expression, although study of both forms in a single experiment is still required.

Chemokines

Chemokines are a superfamily of approximately 40 different small secreted cytokines that direct the migration of immune cells to infection sites. Their activity is coordinated by binding to G-protein-linked receptors with seven transmembrane domains. Four distinct subgroups make up the chemokine superfamily. These are designated as CXC (or a), CC (or b), C (or g) and CX_3C (or d), which are defined by the arrangement of the first two cysteine residues within their peptide structure. The CC subfamily can be further subdivided according to the total number of cysteine residues, as some members of this group contain four cysteines whilst the remainder possesses six (and are known as the C6-b group). Similarly, the CXC subfamily contains two subgroups based on whether or not the first two cysteines are preceded by a Glu-Leu-Arg (ELR) motif associated with specificity to neutrophils.

Interleukin 8

An important chemokine related to the pro-inflammatory process is CXCL-8, also called interleukin 8, this chemokine is a member of the CXC chemokine subfamily and attracts neutrophils, T lymphocytes and basophils in vitro, but not macrophages or monocytes. Many cell-types, including macrophages, produce IL-8 in response to a variety of stimuli (LPS, cytokines and viruses). The neutrophil-attracting ability of IL-8 can be attributed to the presence of the ELR motif adjacent to the CXC motifs at its N-terminus, presumably by affecting its binding to specific receptors. In contrast, CXC chemokines lack an ELR motif and specifically attract lymphocytes but not neutrophils. The biological effects of IL-8 on neutrophils include increased cytosolic calcium levels, respiratory burst, a change in neutrophil shape and chemotaxis.

The fish IL-8 has been found in flounder, trout, catfish, and lamprey. In vitro stimulation of a trout macrophage cell line (RTS-11) or in vivo intraperitoneal challenge with either LPS or poly I:C did result in clear up-regulation of IL-8 expression. Moreover, induction of IL-8 expression in primary cultures of rainbow trout leukocytes stimulated for 24 hours with LPS and TNFα confirms that this fish chemokine is associated with inflammatory response, as has been suggested in mammals. Interestingly, the ELR motif associated with the neutrophil-attracting ability is absent from the lamprey molecule and it is similar in flounder, where CXCL8 also lacks the ELR motif and appears to be regulated by a bacterial mechanism, since its transcript has only been detected in the major immune organs (spleen and head kidney) of an LPS stimulated flounder. The case of the trout is different, although there is also no ELR preceding the CXC motif; it has a very similar motif (DLR) in this position. The human CXCL8 molecule, where the ELR motif has been mutated to DLR, retains neutrophil-attracting ability, albeit at lower potency. Consequently, it is possible that the trout molecule has similar chemotactic activity to that of mammalian CXCL8.

The Interleukin 2 Family

The IL-2 subfamily of cytokines signals via the common gamma chain (gC or CD132), a member of the type I cytokine receptor family expressed in most leucocytes. These cytokines in mammals include IL-2, IL-4, IL-7, IL-9, IL-15 and IL-21. IL-2, IL-4, IL-9 and IL-21 are all cytokines released from The cells, which affect their responses, whilst IL-7 and IL-15 are particularly important for the maintenance of T cell memory. To date molecules with homology to all of these have been found in fish, except IL-9.

Interleukin 2

Interleukin-2 (IL-2 is an important immunomodulatory cytokine that primarily promotes proliferation, activation and differentiation of T cells. IL-2, initially known as T-cell growth factor (TCGF), is synthesized and secreted mainly by Th1 cells that have been activated by stimulation by certain mitogens or by interaction of the T-cell receptor with the antigen/MHC complex on the surface of antigen-presenting cells. Although CD4 T cells are the major source of IL-2 production in response to TCR stimulation, transient induction of IL-2 mRNA and production of the protein has been detected in murine dendritic cells activated by gram-negative bacteria. IL-2 can also be produced by B cells in certain situations. The produced IL-2 promotes the expansion and survival of activated T cells and is also required for the activation of natural killer (NK) cells and for immunoglobulin (Ig) synthesis by B cells.

The IL-2 gene has been detected only recently in fish by analysis of the fugu genome sequence, which also identified IL-21 as a neighbouring gene, as in mammals, providing the first direct evidence for the existence of a true IL-2 homologue in bony fish. The gene has a 4 exon/3 intron organisation, as in mammals, and showed no constitutive expression in a range of tissues examined. However, injection of Fugu with poly I:C induced expression of IL-2 in the gut and gills. Moreover, IL-2 could be induced in head kidney cell cultures stimulated with PHA, and in T-cell enriched cultures isolated from PBL when stimulated with B7-H3 or B7- H4 Ig fusions proteins in the presence of PHA. IL-2 has since been cloned in rainbow trout. The trout IL-2 was significantly up-regulated in head kidney leucocytes by the T cell mitogen PHA and in classical mixed leucocyte reactions and in vivo following infection with bacteria (Y. ruckeri) or the parasite Tetracapsuloides bryosalmonae. More importantly, the recombinant trout IL-2 produced in Escherichia coli was shown to induce expression of two transcription factors (STAT5 and Blimp-1) known to be involved in IL-2 signalling in mammals, as well as interferon-g (IFNγ) and IL-2 itself, and a CXC chemokine known to be induced by IFNγ, termed a IFNγ-inducible protein (γIP).

Interleukin 4

Interleukin-4 IL-4 is a pleiotropic cytokine produced by T cells, mast cells, and basophils and is known to regulate an array of functions in B cells, T cells, and macrophages, hematopoietic and non-hematopoietic cells. IL-4 serves as a key cytokine in driving Th2 differentiation and mediating humoral immunity, allergic responses and certain autoimmune diseases. The IL-4 gene is conserved evolutionarily in the animal kingdom and has been isolated from various animals including humans, mice and bovines, in which the IL-4 locus has been mapped in a region adjacent to those of IL-5 and IL-13 on the same chromosome.

Teleost fish have two genes of the IL-4/13 family, IL-4/13A and IL-4/13B, which are situated on separate chromosomes in regions that duplicated during the fish-specific whole genome duplication (FS-WGD) around 350 million years ago. A few IL-4-like genes have been found in fish to date. The first was discovered by searching the Tetraodon nigroviridis genome. In this work, IL-4 was constitutively expressed in head kidney, spleen, liver, brain, gill, muscle and heart. The ubiquitous expression of IL-4 is consistent with a postulated role in immune cytokines regulation. Stimulating the fish with a mixed stimulant containing ConA, PHA and PMA significantly up-regulated the expression of IL-4, which suggests that IL-4 is involved in the immune inflammatory responses triggered by mitogens, as in mammals, where it has been observed that this mitogen increases IL-4 expression. However, the homology (amino acid identity) of this molecule was very low [12–15%), making it difficult to be sure it is an IL-4 homologue, although clearly related to Th2-type cytokines. In fugu, T cell enriched PBL was found to express more IL-4/13A and IL-4/13B after stimulation with recombinant B7 molecules. In zebrafish a recombinant IL4/13B was shown to increase the number of IgT-positive and CD209-positive cells in blood, and in zebrafish spleen the expression of IL-4/13B and transcription factor related to Th2 immune response as GATA-3, and STAT6 was simultaneously enhanced after PHA stimulation. The IL-4/13A gene was identified in trout and salmon, where the tissue distributions of salmonid IL-4/13A and GATA-3 expression were compared to the expression of IL-4, IL-13, and GATA-3 in mice. High levels of these transcripts were found in both salmonid and murine thymus, while constitutive IL-4/13A richness of skin and respiratory tissue was found in salmonids but not in mice. Experiments with isolated cells from gill and pronephros (head kidney) indicated that trout IL-4/13A is mainly expressed by surface IgM-negative cells, readily inducible by PHA but not by poly I : C, and regulated differently from the Th1 cytokine IFNγ gene. In mammals, IL-5 is also considered a Th-2 type cytokine and along with IL-3 and GM-CSF it signals through receptors with a common γ-chain (γC). None of these cytokines have been discovered in fish to date.

Interleukin 7

The cytokine IL-7 plays several important roles during lymphocyte development, survival, and homeostatic proliferation. It is produced by many different stromal cell types, including epithelial cells of the thymus and the intestine. There is only one report on IL-7 in fish, for the fugu molecule that was discovered using a gene synteny approach by searching with the mammalian IL-7 gene neighbours C8orf70 and PKIA. Fugu IL-7 shows constitutive expression in head kidney, spleen, liver, intestine, gill and muscle, with expression shown to increase in head kidney cultures stimulated with LPS, poly I : C or PHA.

Interleukin 15

The central action of IL-15 cytokine is on T-cells, dendritic cells and NK cells. IL- 15 is an important regulator of the innate immune response to infection and autoimmune disease conditions. This gene shares activities with IL-2 and utilizes IL-2R β and γ units.

Two genes with homology to IL-15 have been discovered in fish. One shows similar gene organisation and synteny to mammalian and chicken IL-15, and has been termed IL-15. The second gene, which has a 4-exon structure and is in a different genome location, has been termed IL-15-like. They show differential expression patterns in terms of the tissues where constitutive expression is apparent and in terms of inducibility in PBL, with IL-15L being refractory to induction. Two alternative splice variants of IL-15L (IL-15La and IL-15Lb) have also been described. Trout IL-15, which has subsequently been cloned and sequenced, was strongly induced by rIFNγ in two trout cell lines (RTS-11 and RTG-2). rIL-15 could up-regulate IFNγ expression in splenic leucocytes, suggesting a positive feedback loop exists in fish between these two cytokines. Interestingly, unstimulated head kidney leucocytes were not responsive to rIL-15, at least in terms of the IFNγ expression level.

Interleukin 21

Interleukin 21 (IL-21) is a newly recognized member of IL-2 cytokine family that utilizes the common γ-chain receptor subunit for signal transduction. In humans and other mammals, IL-21 is produced by both Th1 and Th2 cells. IL-21 has pleiotropic effects on both innate and adaptive immune responses and can act on CD4+ and CD8+ T cells, B cells, NK cells, dendritic cells (DC), myeloid cells, and other tissue cells. IL-21 enhances the proliferation of anti–CD3-stimulated T cells and acts in concert with other γc cytokines to enhance the growth of CD4+ T cells. IL-21–producing CD4+ T cells exhibit a stable phenotype of IL-21 production in the presence of IL-6 but retain the potential to produce IL-4 under Th2-polarizing conditions and IL-17A under Th17-polarizing conditions. IL-21 stimulates CD8+ T cell proliferation and synergizes with IL-15 in promoting CD8+ T cell expansion in vitro and their antitumor effects in vivo. B cells that encounter IL-21 in the context of Ag-specific (BCR) stimulation and T cell co/stimulation undergo class-switch recombination and differentiate into Ab-producing plasma cells. In contrast, B cells encountering IL-21 during nonspecific TLR stimulation or without proper T cell help undergoes apoptosis.

Since its discovery in fugu as a gene neighbour of IL-2, IL-21 has been reported in tetraodon and rainbow trout. Fugu IL-21 shows low constitutive expression. However, stimulation of isolated kidney leucocytes with PHA induced IL-21 expression. IL-21 was also up-regulated at mucosal sites as gill and gut when fish were injected with LPS or poly I : C. Similarly, in tetraodon IL/21 expression is low but detectable in the gut, gonad and gills of healthy fish, and is induced in the kidney, spleen and skin following LPS injection. In trout IL-21 expression is highest in gills and intestine, and is induced in vivo by bacterial (Y. ruckeri) and viral (VHSV) infection. Relative to IL-2, induction of IL-21 expression in head kidney cells appears more rapidly but has shorter duration after stimulation. The trout rIL-21 has also been produced and shown to increase the expression of IL-10, IL-22 and IFNγ, and to a lesser extent IL-21, and to maintain the expression levels of key lymphocyte markers in primary cultures. Thus, IL-21 may act as a survival factor for fish T and B cells.

The Interleukin 10 Family

Interleukin-IL-10 is an anti-inflammatory cytokine and a member of the class II cytokine family that also includes IL-19, IL-20, IL-22, IL-24, IL-26 and the interferons. Although the predicted helical structure of these homodimeric molecules is conserved, certain receptor-binding residues are variable and define the interaction with specific heterodimers of different type-2 cytokine receptors. This leads to diverse biological effects through the activation of signal transducer and activator of transcription (STAT) factors.

Interleukin 10

Interleukin-10 (IL-10) was discovered initially as an inhibitory factor for the production of Th1 cytokines. Subsequently, pleiotropic inhibitory and stimulatory effects on various types of blood cells were described for IL-10, including its role as a survival and differentiation factor for B cells. IL-10, which is produced by activated monocytes, T cells and other cell types like keratinocytes, appears to be a crucial factor for at least some forms of peripheral tolerance and a major suppressor of the immune response and inflammation. The inhibitory function of IL-10 is mediated by the induction of regulatory T cells.

IL-10 was discovered in fish by searching the fugu genome. The translation showed 42–45% similarity to mammalian molecules with very low constitutive expression in tissues. IL-10 has since been cloned in several other fish species including carp zebrafish, rainbow trout, sea bass and cod. Such studies have shown that IL-10 expression can be increased by LPS stimulation, by bacterial infection, by bath administration of immunostimulants and by IPNV infection which may be associated with mechanisms of immune evasion.

Interleukin 20 (IL-20Like)

In mammals, IL-20 was discovered as a new member of the IL-10 family of cytokines. IL-20 shares the highest amino-acid sequence identity with IL-10, IL-24 and IL-19. It is secreted by immune cells and activated epithelial cells like keratinocytes. A high expression of the corresponding IL-20 receptor chains has been detected on epithelial cells. In terms of function, IL-20 might therefore mediate crosstalk between epithelial cells and tissue-infiltrating immune cells under inflammatory conditions.

In fish, the gene of IL-20 has been described in putterfish , zebrafish and trout. In the latter work, the IL-20 gene, called IL-20-like (IL-20L) has been described as having a high level of expression in immune related tissues and in the brain, suggesting an important role of the fish IL-20L molecule in both the immune and nervous systems. Although the exact cell types expressing IL-20L have yet to be defined, macrophages express IL-20L. Moreover, IL-20L expression in the macrophage cell line RTS-11 is modulated by pro-inflammatory cytokines, signalling pathway activators, microbial mimics and the immuno-suppressor dexamethasone. These data suggest that trout IL-20L plays an important role in the cytokine network. The increased expression of IL-20L was only detected at late stages (4–24 h) of LPS stimulation in RTS-11 cells and in spleen 24–72 h after infection with Yersinia ruckeri, which suggests that the increased expression of IL-20L by LPS and infection is via the rapid increase of pro-inflammatory cytokines (e.g., IL-1β) and other factors known to occur.

Interleukin 22/26

In mammals, interleukin-22 is secreted by Th17 cells, as well as by a subset of NK cells, designated as NK22; and even by some Th1 cells. Studies have suggested there is a distinct Th22 cell lineage. Many of the same cytokines that induce differentiation and proliferation of IL-17-producing cells also lead to the secretion of IL-22 by Th17 cells, NK22 cells, and putative Th22 cells, including IL-6, IL-23, IL- 1β, TGF-β, and TNFα. IL-17 and IL-22 are therefore frequently produced together in response to infections. Interleukin-22 interacts with a heterodimeric receptor, IL-10R2/IL-22R1, which is expressed on a variety of non-lymphoid cells, especially epithelial cells. Ligation of this receptor leads to both protective and detrimental effects. In synergy with IL-17, IL-22 induces pro-inflammatory cytokines in human bronchial epithelial cells against Klebsiella pneumoniae infection and in colonic myofibroblasts. Independently or in synergy with IL-17, IL-22 acts in defence against intestinal infection of mice with Citrobacter rodentium. Moreover, IL-22 has been implicated in intestinal homeostasis keeping commensal bacteria contained in anatomical niches, which is key to our symbiotic relationship and normal intestinal physiology. However, the mechanisms that restrict colonization to specific niches are unclear.

IL-26 can be produced by primary T cells, NK cells and T cell clones following stimulation with specific antigen or mitogenic lectins. IL-26 was initially shown by several groups to be co-expressed with IL-22. IL-26 is co-expressed with IFNγ and IL-22 by human Th1 clones, but not by Th2 clones. It was subsequently found that IL-26 is co-expressed with IL-17 and IL-22 by Th17 cells, an important subset of CD4+ T-helper cells that are distinct from Th1 and Th2 cells. More recently, a novel subset of CD56+ NKp44+ NK cells was identified that co-expresses IL-22 and IL-26, especially following treatment with IL-23. Furthermore, a different subset of immature NK cells was described that do not express CD56 or NKp44 but do express CD117 and CD161 and constitutively express IL-22 and IL-26.

The mechanisms that regulate transcription of the human IL-26 gene are so far largely undefined. It is possible and perhaps likely that expression of the IL-26 gene is induced in an IL-23-dependent manner because IL-23 is known to induce differentiation of Th17 cells, and IL-23 amplifies expression of IL-17 and IL-22 by Th17 cells.

In fish, the IFNγ locus was discovered using a gene synteny approach, and was first reported for fugu. It contained a homologue of IL-22/26 that later studies of the zebrafish genome revealed to be two genes, one with clear homology to IL-22 and one with somewhat less clear homology to IL-26. The IL-22 gene was expressed constitutively in intestine and gills in all the treated and non-treated tissues. The gene was also expressed in kidney and spleen in LPS and PolyI:C-treated tissues, respectively, while IL-26 was expressed only in intestine treated with PolyI:C without expression. IL-22 expression has been correlated with disease resistance in haddock vaccinated against V. anguillarum, with a strong constitutive expression in gills in vaccinated fish but not in control fish 24 hours post bath challenge, resulting in complete protection in fish vaccinated. Moreover, IL-22, a cytokine released by Th-17 cells in mammals, is also interesting, and such responses are thought to be crucial for protection against extracellular microbes and at mucosal sites. This coupled with the recent discovery of novel gill-associated immune tissue in fish may provide a clue to a potential mechanism of resistance elicited by the V. anguillarum vaccination.

The Interleukin 17 Family

Interleukin-17 and a related family of genes are known to have pro-inflammatory actions and are associated with diseases. After the discovery of the human IL- 17 gene, five cellular paralogs of IL-17 were identified, namely IL-17B, C, D, E and F. These paralogs, identified by ESTs, genomics and proteomic databases, share identities of 20–50% with IL-17A gene. Human IL-17 A and F are present in tandem in opposite transcriptional orientation on the same chromosome 6p12, while IL-17B (Chr 5q24), IL-17C (Chr 16q24), IL-17D (Chr 13q11) and IL-17E (Chr 14q11) are dispersed. The structural similarities lead to the classification of IL-17 A, B, C, D, E, and F genes to a larger IL-17 sub-family. Several IL-17 family members have been discovered in teleost fish, but homology to mammalian genes has not always been easy to assign. Two IL-17A or F homologue genes (IL-17A/F) have been found on the same chromosome. However, it has been difficult to determine which gene codes IL-17A and F. This gene in zebrafish was named IL-17A/F1 and 2. Furthermore, another IL-17A or F homologue gene (IL-17A/F3) has been found in zebrafish localized on a chromosome different from that of IL-17A/F1 and 2. In addition to those in zebrafish, IL-17A or F homologue genes have been found in rainbow trout, Atlantic salmon, pufferfish (IL-17A/F1, 2 and 3), and medaka (IL-17A/F1, 2 and 3).

The tissue distribution of the fugu IL-17 gene family also differs. In particular, IL-17 family genes are highly expressed in the head kidney and gills. Moreover, expression of IL-17 family genes is significantly up-regulated in the lipopolysaccharide-stimulated head kidney, suggesting that Fugu IL-17 family members are involved in inflammatory responses. In Atlantic salmon IL-17D expression is widely distributed in tissues, with the highest levels of expression in testis, ovary and skin. Infection with A. salmonicida by injection increases IL-17D expression levels in the head kidney (but not the spleen) in a time-dependent manner. Skin and kidney showed an increased IL-17D expression level in fish given a cohabitation challenge with A. salmonicida. The two trout IL-17C genes show some degree of differential expression within tissues, with IL-17C1 being more dominant in the gills and skin, whilst IL- 17C2 is more dominant in the spleen, head kidney and brain. Expression of both genes increases significantly with bacterial infection, although the increased expression of IL-17C2 is greater in terms of fold change. Similarly, both genes could be up-regulated in the trout RTS-11 cell line by LPS, poly I : C, calcium ionophore and rIL-1β, with IL-17C2 showing higher fold increases in all cases.

Interleukin 12

IL-12 is a heterodimeric cytokine composed of p35 and p40 subunits. It can mediate a number of different activities, including stimulation of IFNγ secretion from resting lymphocytes, NK cell stimulation and cytolytic T cell maturation. Perhaps most crucially, IL-12 also affects the progression of uncommitted T cells to either the Th1 lineage, which in general is characterized by secretion of lymphokines associated with cell-mediated rather than humoral immunity.

The p35 and p40 subunits were discovered in fish by analysis of the fugu genome. The p35 locus is quite well conserved, with Schip1 being the immediate neighbour in all cases. This association has allowed p35 to be cloned by gene walking from Schip1 from fish species for which no genome sequence is available. The p40 subunit in fugu is constitutively expressed in all the tissues examined, except muscle, and no increases in expression were seen 3 h after injection with poly I:C or LPS. This constitutive and broad expression distribution of the p40 subunit suggests that it may be expressed in most cell types. The expression of the p35 subunit is more limited in its tissue expression and is induced after injection with poly I : C in the head kidney and the spleen, but not after injection with LPS. These results show that there are differences from the mammalian data in fugu IL-12 subunit expression. Further investigation will be required to show whether this is unique to fugu, if IL-12 is involved more in antiviral defence in fish and if the two subunits are regulated differently from their regulation in the mammalian system.

Transforming Growth Factor β (TGF-β)

TGF-β is a pleiotropic cytokine that regulates cell development, proliferation, differentiation, migration, and survival in various leukocyte lineages including lymphocytes, dendritic cells, NK cells, macrophages and granulocytes. In the mammalian immune system, TGF-β1 is a well-known suppressive cytokine and its dominant role is to maintain immune tolerance and suppress autoimmunity. The potent immunosuppressive effects of TGF-β1 are mediated predominantly through its multiple effects on T cells: TGF-β1 suppress Th1 and Th2 cell proliferation, while it promotes T regulatory cell generation by inducing Foxp3 expression. On the other hand, TGF-β also promotes immune responses by inducing the generation of Th17 cells. Therefore, the regulatory roles of TGF-β as a positive or negative control device in immunity are widely acknowledged in mammals.

In teleost, despite the lack of extensive investigation on the functional role of TGF-β, some recent studies have revealed that TGF-β1 also exterts powerful immune depressing effects on activated leukocytes, as it does in mammals. For instance, TGF-β1 significantly blocks TNFα-induced activation of macrophage in goldfish and common carp, but induces the proliferation of the goldfish fibroblast cell line CCL71. In grass carp, TGF-β1 down-regulates LPS/PHA-stimulated the proliferation of peripheral blood lymphocyte by contrast with the stimulatory effect of TGF-b1 alone in the same cells. In red sea bream, similar phenomenon was observed during leukocyte migration under TGF-β1 treatment, with or without LPS challenges. These findings not only define TGF-β1 as an immune regulator in teleost, but also indicate that TGF-β1 may have retained similar functions in immunity during the evolution of vertebrates.

Interferons

Interferons genes are involved in mediating cellular resistance against viral pathogens and modulating innate and adaptive immune systems. Broadly, IFNs are classified into two main groups called type I and type II. Type I IFN includes the classical IFNα/β, which is induced by viruses in most cells, whereas type II IFN is only composed of a single gene called IFNγ and is produced by NK cells (NK cells) and T lymphocytes in response to interleukin-12 (IL-12), IL-18, mitogens or antigens. Structurally both IFN types belong to the class II a-helical cytokine family, but have different 3-dimensional structures and bind to different receptors.

Two IFNs (IFNα1 and IFNα2) have been cloned from Atlantic salmon and characterized with respect to sequence, gene structure, promoter, antiviral activity and induction of ISGs. Salmon IFNα1 induces both Mx and ISG15 proteins in TO cells and thus has properties similar to mammalian IFNα/β and IFNλ. Furthermore, salmon IFNα1 induces potent antiviral activity against the IPNV in vitro, but this protection has not been observed in vivo, despite a high level of expression of IFNα detected in spleen and head kidney of Atlantic salmon challenged intraperitoneally with IPNV.

At least three type I IFNs have been discovered in rainbow trout. The IFN1 (rtIFN1) and rtIFN2 show high sequence similar to Atlantic salmon IFNα1 and IFNα2, which contains two cysteines. On the other hand, rtIFN3 contains four cysteines, which further confirms the relationship between mammalian IFNα and fish IFNs. Recombinant rtIFN1 and rtIFN2 have both been shown to up-regulate expression of Mx and inhibit VHSV replication in RTG-2 cells. In contrast, recombinant rtIFN3 has been found to be a poor inducer of Mx and antiviral activity. Interestingly, the three rtIFNs show differential expression in cells and tissues. This suggests that the three trout IFNs have different functions in the immune system of fish.

IFNγ has been identified in several fish species, including rainbow trout and Atlantic salmon. In contrast to the type I IFNs, fish and mammalian IFNγ are similar in exon/intron structure and display gene synteny. However, some fish species also possess a second IFNγ subtype named IFN gamma rel, which is quite different from the classical IFNγ. Rainbow trout and carp IFNγ have several functional properties in common with mammalian IFNγ, including the ability to enhance respiratory burst activity, nitric oxide production, and phagocytosis of bacteria in macrophages. Far less is known about the antiviral properties of fish IFNγ. However, it has been reported that it induces antiviral activity against both IPNV and the Salmon Alpha Virus (SAV) in salmon cell lines.

Osmoregulation

Osmoregulation is the process of maintaining an internal balance of salt and water in a fish's body. A fish is, after all, a collection of fluids floating in a fluid environment, with only a thin skin to separate the two.

There is always a difference between the salinity of a fish's environment and the inside of its body, whether the fish is freshwater or marine. Since the fish's skin is so thin, especially around places like the gills, external water constantly tries to invade the fish's body by osmosis and diffusion.

The two sides (inside and out) of a fish's membrane skin have different concentrations of salt and water. Nature always tries to maintain a balance on both sides, so salt ions will move through the semi-permeable membrane towards the weaker salt solution (by diffusion), while the water molecules take the opposite route (by osmosis) and try to dilute the stronger salt solution.

Regardless of the salinity of their external environment, fish use osmoregulation to fight the processes of diffusion and osmosis and maintain the internal balance of salt and water essential to their efficiency and survival.

The ability of some fish (e.g., .salmon) to regulate in both environments during migration is of great interest.

In fishes the kidneys play an important role in osmoregulation, but major portion of the osmoregulatory functions are carried out by other organs such as the gills, the integument and even the intestine. Osmoregulation may be defined as "the ability to maintain a suitable internal environment in the face of osmotic stress".

As a consequence there is always difference between the optimal intracellular and extracellular concentrations of ions. In the fish body, number of mechanisms takes place to solve osmotic problems and regulate the difference.

The most common ones are:

1. Between intracellular and extracellular compartment.

2. Between extracellular compartment and the external environment. Both are collectively called 'osmoregulatory mechanisms', a term coined by Rudolf Hober.

Problems of Osmoregulation

Generally fish lives in an osmotic steady state in spite of frequent variations in osmotic balance. That is, on the average, the input and output being equal over a long period sum up to zero.

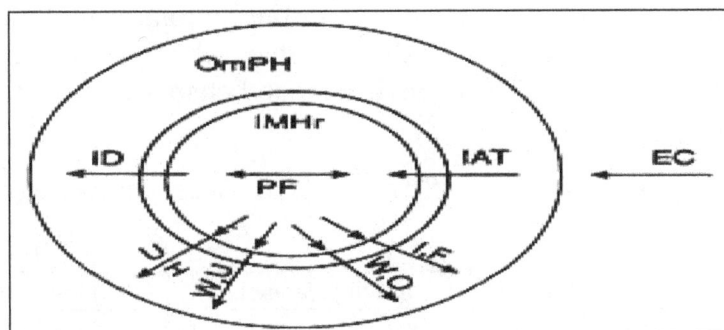

Principal process of osmoregulation in freshwater fishes. EC, ecological condition;
I, ion; IAT, ion active transfer; ID, ion diffusion; ITIFir, inner medium hypertonic; o, osmosis;
OmPH, outer medium hypotonic; PF, physiological factor; U, urine; W, water.

The osmotic exchanges that take place between the fish and its environment may be of two types:

1. Obligatory Exchange: It occurs usually in response to physical factors over which animal has little or no physiological control and

2. Regulatory Exchange: These are the exchanges which are physiologically well controlled and help in the maintenance of internal homeostasis.

Factors affecting Obligatory Exchanges

Gradient between the Extracellular Compartment and the Environment

The greater the ionic difference between the body fluid and external medium, the greater the tendency for net diffusion to low concentrations. Thus, a bony fish in a sea water is affected by the problem of losing water into the hypertonic sea water.

Surface/Volume Ratio

Generally the animal with small body size desiccates (or hydrates) more rapidly than a larger animal of the same shape.

Permeability of the Gills

Fish gills are necessarily permeable to water and solutes as they are the main site of exchange of oxygen and carbon dioxide between the blood and the water. Active transport of salts also takes place in the gills. Euryhaline fishes (who have tolerance of wide range of osmolarity) are well adapted to saline water by reduced permeability to water.

Feeding

Fishes take water and solute along with the feeding. A gill takes high quantity of salt than water at the time of feeding on seashore invertebrates, these fishes, therefore, must have some special device to excrete excess of salt. However, a freshwater fish ingests large amount of water than salt and thus needs special means of salt conservation.

Osmoregulators and Osmoconfirmers

Osmoregulators are those animals who can maintain the internal osmolarity different from the medium in which they live. The fishes, except the hagfish which migrates between fresh and saline waters, the changing osmotic stress due to environmental changes is overcome with the help of endocrine mechanism.

Table: Approximate composition of extracellular fluids of teleostean fishes (concentration in millimoles per litre of water).

Species	Habitat	Milli	Na^+ mole	K^+	Ca^{++}	Mg^{++}	Cl^-	SO_4^{--}
Paralichthys(founder)	Sea-water	337	180	4	3	1	160	0..2
Carrasius	Fresh-Water	293	142	2	6	3	107	-

Osmoconfirmers are those animals who are unable to control osmotic state of their body fluids but confirms to the osmolarity of the ambient medium. Majority of fishes either live in freshwater or in salt water (a few live in brackish water).

Due to various physiological processes, metabolic wastes are removed from the body in vertebrates by gut, skin and kidneys. But in fishes and aquatic animals their gills and oral membranes are permeable both to water and salts in marine environment, salt is more in water against the salt inside the body fluid, hence water moves out due to the process of 'osmosis'.

The 'osmosis' may be defined as "if two solutions of different concentrations are separated by a semipermeable membrane, the solvent from the less concentrated part will move through the membrane into more concentrated solution." Hence to compensate the loss of water marine fishes drink water.

The salt will enter the body due to concentration gradient and so salt will be more inside the body. On the other hand, in freshwater fishes, the salt will go out to the environment as the salt concentration will be more inside the body fluid. The water will move inside the body due to osmosis through partially permeable membrane.

This means solvent will pass into more concentrated solution, but solute will also pass in the opposite direction. There will be, however, a difference in the rate dependent upon the relative permeability for two types of molecules usually solvent pass rapidly.

Osmoregulation in Freshwater Fishes

The body fluid of freshwater fishes is generally hyperosmotic to their aqueous medium. Thus they are posed with two types of osmoregulatory problems.

1. Because of hyperosmotic body fluid they are subjected to swelling by movement of water into their body owing to osmotic gradient.

2. Since the surrounding medium has low salt concentration, they are faced with disappearance of their body salts by continual loss to the environment. Thus freshwater fishes must prevent net gain of water and net loss of salts. Net intake of water is prevented by kidney as it produces a dilute, more copious (i.e., plantiful hence dilute) urine.

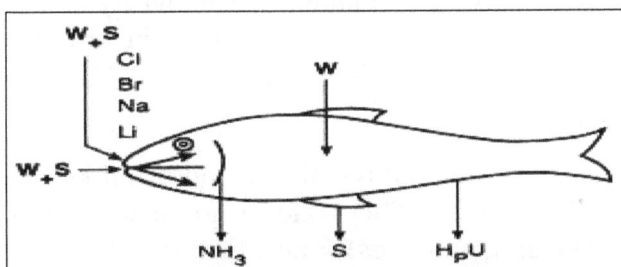

Osmoregulatory in flow and outflow of salts and water
in a fresh water fish. HpU, hypotonic urine; S, salt;
W, water, W+S, water and salt.

The useful salts are largely retained by reabsorption into the blood in the tubules of kidney, and a dilute urine is excreted. Although some salts are also removed along with urine which creates torrential loss of some biologically; important salts such as KCl, NaCl, $CaCl_2$ and $MgCl_2$ which are replaced in various parts.

Freshwater fishes have remarkable capacity to extract Na^+ and Cl^- through their gills from surrounding water having less than 1 m M/L NaCl, even though the plasma concentration of the salt exceeds 100 m M/L NaCl.

Thus NaCl actively transported in the gills against a concentration gradient in excess of 100 times. In these fishes the salt loss and water uptake are reduced by the integument considerable with low permeability or impermeability to both water and salt also by not drinking the water.

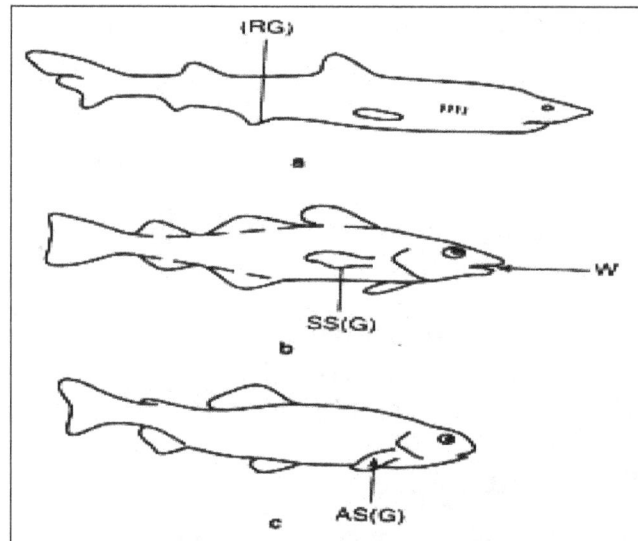

In figure, exchange of water and salt in some fishes. (a) Marine elasmobranch does not drink water and has isotonic urine. (b) Marine teleost drinks water and has isotonic urine. (c) Freshwater teleost drinks no water and has strongly hypotonic urine; AS(C), absorbed salt with gill; HrNaCI (RC), hypertonic NaCI from rectal gland; WC). Secretes salts from gill; W, water.

Osmoregulation in Marine Water Fishes

In marine fishes, the concentration of body fluid and marine water is almost similar. Therefore, they do not require much energy for maintenance of osmolarity of their body fluid. The classic example is hagfish, Myxine whose plasma is iso-osmotic to the environment. Hagfish maintains the concentration of Ca^{++}, Mg^{++} and SO_4 significantly lower and Na^+ and CI higher in comparison to sea water.

Other marine water fishes such as sharks, rays, skates and primitive coelacanth, Latimaria, have plasma which is iso-osmotic to sea water. They differ from the hagfish in having capacity to maintain very lower electrolyte (i.e., inorganic ions) concentrations.

They also have difference with organic osmolytes like urea and trim-ethylamine oxide. Kidneys of coelacanth and elasmobranchs excrete excess of inorganic salts such as NaCl.

Also rectal gland located at the end of alimentary canal takes part in the excretion of NaCl. Modern bony fishes (marine teleost) have the body fluid hypotonic to sea water, so they have tendency to lose water to the surroundings particularly from gill via epithelium. The lost volume of water is replaced by drinking salt water.

About 70—80% sea water containing NaCl and KCl enters the blood stream by absorption across the intestinal epithelium. However, most of the divalent cations like Ca^{++}, Mg^{++} and SO4 which are left in the gut are finally excreted out.

Excess salts absorbed along with sea water is ultimately received from the blood with the help of gills by the active transport of Na^+Cl^- sometimes K^+ and eliminated into the sea water. However, divalent ions are secreted into the kidney.

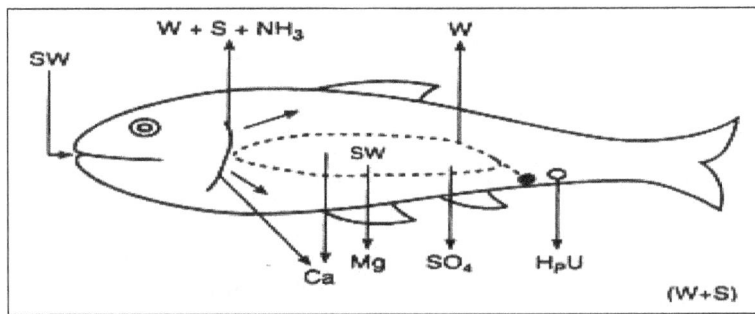

Osmotic regulation in marine bony fishes. HpU, hypotonic urine;
SW, sea water; W+S+ NH$_3$,water, salt and ammonia ; W, water.

Thus, urine is isosmotic to the blood but rich in those salts, particularly Mg^{++}, Ca^{++} and SO$_4^{--}$which are not secreted by the gills. Combined osmotic action of gills and kidney in marine teleost resulted in the net retention of water that is hypotonic both to the ingested water and urine.

By using similar mechanism some teleost species such as the salmon of Pacific Northwest maintain more or less constant plasma osmolarity in-spite of being migratory between marine and freshwater environment.

The fishes can be divided into four groups on strategies of regulation of internal water and total solute concentrations.

Controls of Osmoregulation

The concentration and dilution of urine is controlled by hormones, which affects the rate of renal filtration by changing the blood pressure and thus control the quantity of urine. Hormones also influence the rate of diffusion and absorption across the gill epithelium. Thyroid gland and supra-renal bodies secrete adrenocortical hormones which control osmoregulation in fishes.

Thermoregulation

Thermoregulation is the process by which an organism controls its internal temperature. Fish have many different mechanisms for regulating their temperature. Most fish are ectothermic, using their environmental temperature to manage their body temperature, but some fish are endothermic, having the metabolic ability to internally manage temperature. Poikilothermic fish are ectotherms which have no control over their body temperature; their core body temperature conforms to ambient temperature. Eurythermic fish have evolved to survive in a wide range of environmental temperatures and stenothermic fish have evolved to survive in a narrow range of environmental temperatures.

Thermoregulation is very important for fish because temperature influences the function of many organs and the rate of many metabolic processes. Most fish species have evolved to survive within a specific temperature ranges; outside that range, enzymes can degrade, organs can fail, and the organism can die. Understanding thermoregulation for fish species is particularly important when considering implications for climate change.

Behavioral thermoregulation occurs when a fish actively seeks out areas of water with higher or lower temperature. For instance juveniles of the Bear Lake Sculpin (Cottus extensus) live at the bottom of lakes feeding on the lake floor during the day. Here the temperature is normally around 5 °C, however after dark they stop feeding and rise to the surface where they hang around digesting their food. The temperature near the surface of the lake is about 14 °C. By doing this they are able to digest much more food during the night than they could have done by staying on the lake bottom and so they can grow more quickly. The difference is considerable, at 5 °C it would take them 33 hours, or a day and a half to digest the food they can eat in one day, but at 14 °C it takes them only 4.5 hours to digest the same amount of food, so by coming to the surface at night they are making much more efficient use of their time. And then by returning to the colder water as soon as its food is digested it slows down its metabolism again and thus conserves energy.

Sometimes the strategies that evolve in simple living creatures are truly amazing and almost seem to reflect intelligence. Another example of behavioural thermoregulation is the Sockeye Salmon (Oncorhynchus nerka). Sockeye Salmon are crepuscular feeders, meaning they hunt for food, and therefore feed only at dawn and dusk. During the full day and the full night they rest. Scientists have learned that in Canada, in summer, when the days are long and the nights are short that the salmon adjust their behaviour to maximize their efficiency. After feeding at dusk the salmon rest throughout the night at depths of around 11 metres where the temperature is about 15 °C, however after their dawn feeding they retreat to depths of about 37 metres where the temperature is only about 5 °C. Why? Because the night is short they need to be warm to digest all they have eaten before the next feeding session, but because the days are long they have more time and so can digest their food fully at a lower temperature and in doing so they reduce their metabolic costs.

Behavioural thermoregulation is not uncommon in fish, or in reptiles, amphibians and insects. It is easy to implement and has the advantage of not being permanent, meaning the animal can, in the best environmental circumstances choose to be warm or cool as it needs. However it has the problem that the environment is not always at its best, and even when it is it takes time to warm up enough to allow quick muscular action. Animals that practice physiological thermoregulation have the edge when the ambient temperature is not optimum, which even in the sea is quite often.

Physiological thermoregulation is where the fish controls its core body temperature by means of internal physiological and metabolic activities. This is also how we maintain our body temperatures, but while it is universal among mammals it is rare in fish. It occurs in only a few species, all of which are marine and swim constantly. These include various Tuna, some Mackerels, the Mackerel Sharks (Lamna nasus and L. ditropis) and the Great White Shark (Carcharodon carcharias).

Both sharks and bony fish that maintain an increased body temperature do so by means of a counter-current exchange system whereby blood vessels carrying blood that is hot as a result of muscular activity pass along side, and give up some of their heat to, blood that is going to parts of the body the animal wishes to keep warm. The organ where the heat exchange takes place is called retia mirabilia. A fish may have more than one retia Mirabella; Mackeral Sharks for instance have three, one in the swimming muscles, one in the body cavity near the guts and one around the brain.

As with any other successful adaptation there are many other minor changes that work along side the major one in harmonizing the animals metabolism. For example Albicore Tuna have evolved a unique form of haemoglobin that increases its affinity for O_2 as it gets warmer while its CO_2 affinity increases as it gets cooler. This is the reverse of the normal situation, but it does prevent O2 from being lost from the warming arterial to the cooling renal blood in the retia mirabilia.

The next step up from recycling the heat from muscular effort to heat parts of the body, which obviously only works while the animal is using it muscles (which is why all the species that do so are constant swimmers), is to use purely metabolic processes to generate heat. The only example is the Swordfish (Xiphia gladius). Swordfish have specifically modified eye muscles with an exceptionally high mitochondrial volume. Mitochondria are the organelles wherein food is turned into energy, they are found in every cell of your body. The swordfish does not use these special eye muscles for anything except normal eye movements (which does not generate much heat) and to heat the blood going to its brain. In other words the mitochondria work purely to make heat for the purpose of keeping the precious brain warm, it is not merely recycling heat that was a byproduct of normal muscular activity as are the sharks and the tuna. This is much closer to the mechanisms used by birds and mammals to maintain their body temperatures. Evolution takes a long time in comparison with short life spans, but who can say what fish will be swimming in the sea and oceans in another 50 million years.

Chemical reactions occur more quickly at warmer temperatures, warmer guts digest food more quickly, a warmer heart beats more strongly, warmer muscles contract more strongly and respond more quickly and warmer brains work better. The contraction power of the muscles of Bonito (Sarda chiliensis lineolata) doubles with a 10 °C increase in temperature. All this can give a predator and advantage that makes the difference between life and death. All the fish and sharks that maintain a raised body temperature are active constant hunters. The tuna and mackerels hunt smaller fish and the sharks hunt the tuna and mackerel amongst other things. Endothermic animals need to eat a lot more than ectothermic ones, sometimes as much as 3 or 4 times as much.

Poikilotherm

Poikilothermic fish have no control over their body temperature and their core body temperature can fluctuate broadly. While some ectothermic stenotherms thermoregulate their body temperature by inhabiting constant temperature environments, internal temperature of poikilotherms can widely vary.

Thoughout their lives, Steelhead's internal temperature varies considerably (NPS).

Eurytherm

In contrast to stenotherms, eurythermic fish can function at a wide range of water temperatures. They are often, but not necessarily, ectotherms. Desert Pupfish (Cyprinodon macularius), for example, can function in ambient temperatures ranging from 4 to 45 degrees Celsius. This thermoregulatory strategy requires that organs, enzymes, and metabolic processes can operate at varying environmental temperatures.

Desert Pupfish are eurythermic, surviving in temperatures ranging from function in waters from 4° to 45 °C.

Stenotherm

In contrast to eurytherms, senothermic fish can only function in a narrow range of water temperatures. Brook Trout (Salvelinus fontinalis), for example, function optimally approximately between 13 and 18 degrees Celsius. This thermoregulatory strategy requires that organs, enzymes, and metabolic processes operate in a small temperature band and makes these fish particularly vulnerable to environmental changes.

Brook Trout can only survive in a narrow band of temperatures (FWS).

Ectotherm

Unlike endotherms which can metabolically control their own body temperature, ectotherms rely upon environmental temperatures for thermoregulation. Most fish are ectotherms. Ectothermy can be metabolically more efficient than endothermy because organisms do not have to expend any energy to self-regulate their body temperatures. However, they are at the mercy of their environment more than endotherms because they use ambient water temperature to control their body temperature.

Poikilothermic fish have no control over their body temperature whatsoever. Their core body temperature conforms entirely to ambient temperature and can fluctuate widely. This means that their organs and enzymes need to be capable of functioning at a range of temperatures. As physiological processes have often evolved to operate most efficiently at certain temperatures, ectothermic fish have evolved multiple strategies to maintain optimal thermal habitat. Stenotherms live within narrow environmental temperatures in contrast with eurytherms which can live in a wide range of environmental temperatures.

Endotherm

Unlike ectotherms, which rely upon environmental temperatures, endotherms are able to metabolically control their body temperature. This thermoregulatory strategy is rare among fish but is present in tunas and some sharks, including the Great White Shark (Carcharodon carcharias) and Shortfin Mako Shark (Isurus oxyrinchus). Endothermic tunas and sharks use a network of capillaries in their swimming muscles, the Rete mirabile, as a heat exchanger. Through counter-current exchange, the heat produced through muscle activity is transported by the blood. Through this metabolic process, sharks, for example, can maintain a body temperature of 5 – 14 °C above ambient water temperature. This process is an evolutionary advantage for this long distance, migratory fish, allowing them to travel extensive distances and dive deep while maintaining body temperature, conserving energy, and avoiding thermal shock from changes in water temperature.

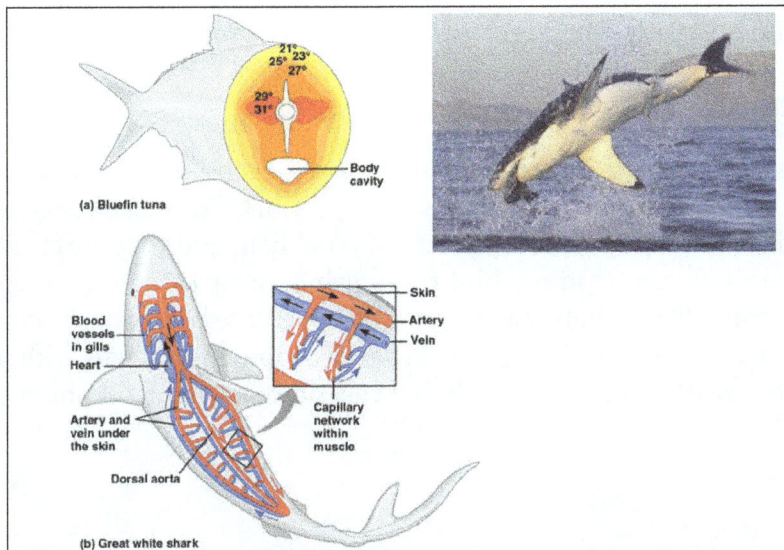

Though rare, some fish are able to internally regulate temperature.

Buoyancy

Fish live in a three-dimensional environment. Instead of just worrying about moving forward, backward and to the sides, fish must also think about moving up and down. This typically requires a change in buoyancy; many fish have the ability to make they lighter to float up or heavier to sink down by using a swim bladder.

For a fish to be buoyant, or float, it must displace less water or the same amount of water as its own body mass. This isn't easy in an underwater environment where the density of the water changes with the depths and different currents. Many fish use swim bladders to help them with quick depth changes. These bladders fill with air to help fish rise or release air so fish can sink, often in conjunction with forward movement. Some fish, however, don't have swim bladders. These fish typically use continuous forward movement along with angled pectoral fins to stay afloat.

Open Swim Bladders

When you see fish swim to the surface and open their mouths to gulp air, many aren't breathing that air; they likely still breathe through their gills. Instead, these fish are filling their swim bladders with air to help maintain buoyancy. These swim bladders are open, meaning there is a direct line from the mouth to the swim bladder. Fish with open swim bladders, typically freshwater fish, tend to live in shallow water or within a few feet of the surface to allow easy access to the air.

Closed Swim Bladders

Closed swim bladders are more efficient for fish that prefer to live deeper underwater. Instead of having a tube from the mouth to the swim bladder, these fish have closed swim bladders that pull air out of connected blood vessels to stay buoyant at lower depths or to rise quickly. To sink, the closed swim bladders release air back into the circulatory system. The air in this system comes from the gills, so it is constantly being replenished to be available when the swim bladder needs it.

Fish with No Bladders

Bony fish, or the ones with hard bones similar to those of land vertebrates, typically have swim bladders. Fish with skeletons made of cartilage, such as sharks, often use another system to maintain buoyancy. Their skeleton is lighter than that of bony fish, and they carry oil in their livers; oil is usually lighter than water, providing a bit of buoyancy. Sharks' main system of staying afloat is continuous swimming. Their stationary pectoral fins create a certain level of buoyancy when combined with forward movement. Because the sharks propel themselves with powerful tails, they can change depths faster than most bony fish that rely on releasing air from or adding air to swim bladders

Locomotion

Swimming is the most economical form of animal locomotion because the body of aquatic animals is supported by water and hence the animals do not have to spend energy to counter gravity. While a squirrel spends 5.43 kcal for walking and a gull spends 1.45 kcal for flying, a salmon spends only 0.39 kcal for swimming one kilometer distance. Almost neutral buoyancy is achieved by the air bladder of bony fishes and by the fatty liver in sharks. Streamlined body of fishes provides least resistance while moving through viscous water.

Based on energy costs, swimming can be classified into 3 groups, namely, sustained, prolonged and burst swimming.

Sustained Swimming

Swimming speed is slow, almost 6-7 body lengths per second and is maintained for long periods. Energy required by muscles is provided by aerobic respiration and since the speed is slow, oxygen debt is not built up as fatigue comes very slowly. This type of locomotion is used for foraging over large areas or for long distance migrations.

Burst Swimming

This type of locomotion is used for escaping predators, chasing a prey or for swimming against currents of water. High speeds of up to 20 body lengths/second are achieved but can be sustained only for short periods. Power is generated by anaerobic respiration in which creatinine phosphate and muscle glycogens are used up. Fatigue comes very rapidly and therefore burst swimming can be sustained for short periods.

Prolonged Swimming

This type of swimming is intermediate between the above two types in speed and energy. Energy is supplied by both aerobic and anaerobic respiration. Prolonged swimming can last up to 3 hours and longer bouts can end up in fatigue. This type of locomotion is used occasionally as the situation demands.

Fish Hydrodynamics

Fishes live in viscous medium where they face two types of drags that must be overcome in order to move forward in water. Fishes are able to move through water causing almost no turbulence:

- Viscous drag: As the fish moves forward viscous drag is created due to friction between the fish body surface and water. Thin and small body faces high viscous drag owing to larger surface area as compared to bulky body. Therefore, larval fishes having larger surface area, experience viscous drag as a major force against them. Hence these tiny creatures must always actively swim since they are unable to glide forward without actively swimming. Mucous coated streamlined body of fishes reduces viscous drag.

- Pressure drag or inertial drag: Displacement of water by fish body creates pressure drag. It is the pressure of water from all sides on fish body. Bulky fishes face high inertial drag as they replace more water as compared to smaller fishes. Pressure drag increases with speed as well as depth. Streamlining reduces pressure drag and fishes keep their body in straight line in order to minimize drag.

Aspect Ratio of Caudal Fin

Aspect ratio is the dorsal to ventral width of caudal fin divided by anterior to posterior length. High aspect ratio gives efficient forward motion as in sharks. Caudal fins of trouts, minnows and perches are flexible and can change aspect ratio according to the needs.

Fishes swim by metachronal contraction of myomeres alternately on either side of axis. Lateral push of the caudal fin on water produces a reactive force on the opposite side at right angle to the axis of body. Reactive force has two components—forward thrust and lateral force. Thrust propels the body forward and overcomes drag while lateral force makes the head yaw from side to side. Push force of the caudal fin is always stronger than any other force.

- Yawing is side to side movement of head created by the lateral reaction force generated by the sideways lashing of the tail fin. Yawing is countered by the use of pectoral fin so that fish can move in straight line.

- Pitching is up and down movement of the head produced by uneven drag on the body or by heterocercal or hypocercal tail fin. Pitching is countered by pectoral fins.

- Rolling is spinning of the body on its anterior-posterior axis. Rolling must be controlled while turning right or left and it is done by the dorsal fin. Bony fishes have foldable dorsal fin that is supported by fin rays, so that rolling can be controlled at will.

There are three types of locomotions in fishes depending on the shape of body.

Anguilliform Locomotion

Eels (Anguilla) and cyclostomes having serpentine body swim by lateral undulation of the entire body that is caused by metachronal rhythm in the contraction of myotomes. This type of swimming is quite efficient at low speeds but consumes a lot of energy since the whole of the body is involved in locomotion.

Carangiform Locomotion

In majority of fishes lateral undulation of body is restricted to the posterior one-third of body. Tail is lashed from side to side in such a way that it always has a backwardly facing component of push and caudal fin increases the area and the force of backward push of tail.

Ostraciform Locomotion

This type of locomotion is found in box fishes and trunk fishes (family Ostraciidae) in which body is not flexible and hence cannot undergo lateral undulation. Therefore, only tail fin propels the body forward.

Locomotion in Sharks and Dogfishes (Pleurotremata)

Sharks and dogfishes have long, streamlined body with a heterocercal tail fin. Pectoral fins are located in front of the centre of gravity that lies just below the dorsal fin. Larger upper lobe of the caudal fin produces lift force on tail due to which head pitches downward. Pitching force is countered by the pectoral fins which also function as elevators. Heterocercal tail fin helps elasmobranchs to swim near the bottom of sea as most of the elasmobranchs are natural bottom dwellers.

Dorsal fins are antirolling devices and they stop rolling of the body while the fishes turn right and left. Pelvic fins in cartilaginous fishes do not contribute to swimming or balancing.

Locomotion in Skates and Rays (Hypotremata)

Like other elasmobranchs, rays also have heterocercal tail fin and two dorsal fins on the tail. But they have a doroventrally flattened body and enlarged pectoral fins are fused on the lateral margins of body. Pectoral fins can produce metachronal contractions and propel the body forward. Rays being dorsoventrally flattened have no problem of rolling and hence dorsal fins are reduced.

Locomotion in Bony Fishes

Majority of bony fishes possess homocercal or diphycercal tail fin that produces a straight forward push on the body to counter viscous as well as pressure drag. Dorsal fin is foldable and can be stretched whenever required. Pectoral fins are placed high and are used as brakes and for turning right and left. Anteriorly placed pelvic fins stop the upward lift of head while braking. Bony fishes also use operculum to eject water to help in quick turning. Swim bladder maintains the fish steady at a given depth.

Locomotion in Flying Fish

Flying fishes, owing to their enlarged pectoral fins can glide in air for considerable distances. Caudal fin is hypocercal with enlarged lower lobe that helps to pull the tail down and keep head upwards while swimming so that they can swim upward rapidly and jump out of water to glide. Even pelvic fin is enlarged to give upward lift to the body.

Locomotion in Sea Horse and Pipe Fish

Sea horse and pipe fish have no fins except the single dorsal fin and hence this fin is used to push the body forward in a vertical position. Tail is prehensile to hold on to the sea weeds and corals where these creatures remain camouflaged and prey upon planktons.

Some bony fishes such as Amia have very long dorsal fin extending to almost the entire length of the back. This fin is capable of undulation to propel the body forward while swimming at slow speed. Similarly, Notopterus and Wallago have very long anal fin which almost continues up to the tail fin and is used to push the body forward.

Trigger fish that can produce fast bursts of speed at short distances, have very high caudal fin to increase the surface area. Dorsal fin and anal fins are also broad and placed posteriorly near the caudal fin to increase the aspect ratio of the posterior region so that a powerful push can be created to propel the body forward.

Forces Acting on a Swimming Fish

The main properties of water as locomotion medium, that have played an important role in the evolution of fish, are its incompressibility and its high density. Since water is an incompressible fluid, any movement executed by an aquatic animal will set the water surrounding it in motion and vice versa. It's density (about 800 times that of air) is sufficiently close to that of the body of marine animals to nearly counterbalance the force of gravity. This has allowed the development of a great variety of swimming propulsors, as weight support is not of primary importance.

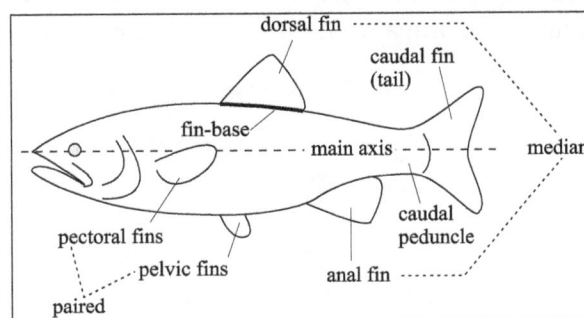

Terminology used in the text to identify
the fins and other features of fish.

To aid in the description of the fish swimming mechanisms, figure illustrates the terminology used to identify morphological features of fish. Median and paired fins can also be characterised as either short-based or long-based, depending on the length of their fin base relative to the overall fish length. The fin dimensions normal and parallel to the water flow are called span and chord respectively.

Swimming involves the transfer of momentum from the fish to the surrounding water (and vice versa). The main momentum transfer mechanisms are via drag, lift and acceleration reaction forces. Swimming drag consists of the following components:

a. Skin friction between the fish and the boundary layer of water (viscous or friction drag). Friction drag arises as a result of the viscosity of water in areas of flow with large velocity gradients. Friction drag depends on the wetted area and swimming speed of the fish, as well as the nature of the boundary layer flow.

b. Pressures formed in pushing water aside for the fish to pass (form drag). Form drag is caused by the distortion of flow around solid bodies and depends on their shape. Most of the fast-cruising fish have well streamlined bodies to significantly reduce form drag.

c. Energy lost in the vortices formed by the caudal and pectoral fins as they generate lift or thrust (vortex or induced drag). Induced drag depends largely on the shape of these fins.

The latter two components are jointly described as pressure drag. Comprehensive overviews of swimming drag (including calculations for the relative importance of individual drag components) and the adaptations that fish have developed to minimise it can be found in.

Like pressure drag, lift forces originate from water viscosity and are caused by assymetries in the flow. As fluid moves past an object, the pattern of flow may be such that the pressure on one lateral side is greater than that on the opposite. Lift is then exerted on the object in a direction perpendicular to the flow direction.

Acceleration reaction is an inertial force, generated by the resistance of the water surrounding a body or an appendage when the velocity of the latter relative to the water is changing. Different formulas are used to estimate acceleration reaction depending on whether the water is accelerating and the object is stationary, or whether the reverse is true. Acceleration reaction is more sensitive to size than is lift or drag velocity and is especially important during periods of unsteady flow and for time-dependent movements.

The forces acting on a swimming fish are weight, buoyancy and hydrodynamic lift in the vertical direction, along with thrust and resistance in the horizontal direction.

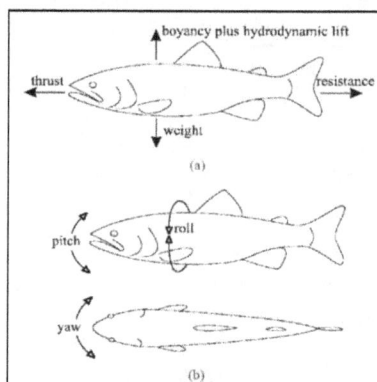

(a) The forces acting on a swimming fish. (b) Pitch, yaw and roll definitions.

For negatively buoyant fish hydrodynamic lift must be generated to supplement buoyancy and balance the vertical forces, ensuring that they do not sink. Many fish achieve this by continually swimming with their pectoral fins extended. However, since induced drag is generated as a side-effect of this technique, the balance between horizontal forces will be disturbed, calling for further adjustments for the fish to maintain a steady swimming speed. The hydrodynamic stability and direction of movement are often considered in terms of pitch, roll and yaw. The swimming speed of fish is often measured in body lengths per second.

For a fish propelling itself at a constant speed the momentum conservation principle requires that the forces and moments acting on it are balanced. Therefore, the total thrust it exerts against the water has to equal the total resistance it encounters moving forward. Pressure drag, lift and acceleration reaction can all contribute to both thrust and resistance. However, since lift generation is associated with the intentional movement of propulsors by fish, it only contributes to resistance for actions such as braking and stabilisation rather than for steady swimming. Additionally, viscous drag always contributes to resistance forces. Finally, body inertia, although not a momentum transfer mechanism, contributes to the water resistance, as it opposes acceleration from rest and tends to maintain motion once begun. The main factors determining the relative contributions of the momentum transfer mechanisms to thrust and resistance are (i) Reynolds number, (ii) reduced frequency and (iii) shape.

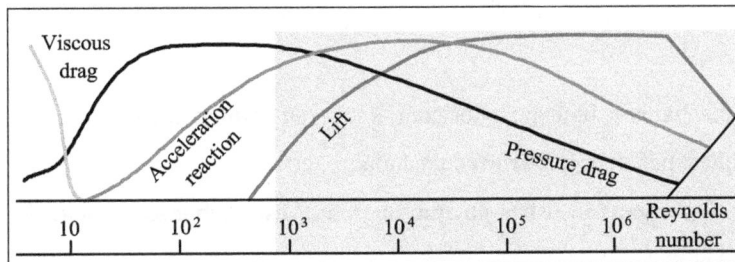

Diagram showing the relative contribution of the momentum transfer mechanisms for swimming vertebrates, as a function of Re. The shaded area corresponds to the range of adult fish swimming.

The Reynolds number (Re) is the ratio of inertial over viscous forces, defined as:

$$Re = \frac{LU}{v}$$

Where, L is a characteristic length (of either the fish body or the propulsor), U is the swimming velocity and v is the kinematic viscosity of water. In the realm of Re typical of adult fish swimming g (i.e. $10^3 < Re < 5.10^6$) inertial forces are dominant and viscous forces are usually neglected. At those Re acceleration reaction, pressure drag and lift mechanisms can all generate effective forces.

The reduced frequency σ indicates the importance of unsteady (time-dependent) effects in the flow and is defined as:

$$\sigma = 2\pi \frac{fL}{U}$$

Where, f is the oscillation frequency, L is the characteristic length and U is the swimming velocity. The reduced frequency essentially compares the time taken for a particle of water to traverse the length of an object with the time taken to complete one movement cycle. It is used as a measure of the relative importance of acceleration reaction to pressure drag and lift forces. For σ < 0.1 the movements considered are reasonably steady and acceleration reaction forces have little effect. For 0.1 < σ < 0.4 all three mechanisms of force generation are important, while for larger values of σ acceleration reaction dominates. In practice, for the great majority of swimming propulsors, the reduced frequency rarely falls below the 0.1 threshold.

Finally, the shape of the swimming fish and the specific propulsor utilised largely affects the magnitude of the force components. The relationship is well documented for steady-state lift and drag forces, but relatively little work has been done on the connection between shape and acceleration reaction.

A common measure of swimming efficiency is Froude efficiency η , defined as:

$$\eta = \frac{\langle T \rangle U}{\langle P \rangle}$$

Where, U is the mean forward velocity of the fish, T is the time-averaged thrust produced and P is the time-averaged power required.

References

- Respiration-fish-mechanism, biology: byjus.com, Retrieved 15 July, 2019

- Respiration-in-fishes: bdfish.org, Retrieved 29 January, 2019

- Accessory-respiratory-organs-in-fishes-phylum-chordata, fish, fisheries: biologydiscussion.com, Retrieved 9 June, 2019

- Blood-circulatory-system-in-fish, anatomy-and-physiology, fish: yourarticlelibrary.com, Retrieved 19 May, 2019

- Digestive-system-in-fishes-with-diagram, anatomy-and-physiology, fish: yourarticlelibrary.com, Retrieved 17 August, 2019

- Types-of-Reproduction-in-fishes: aquafind.com, Retrieved 30 January, 2019

- Growth-of-fishes-with, anatomy-and-physiology, fish: yourarticlelibrary.com, Retrieved 28 March, 2019

- Fish-cytokines-and-immune-response, new-advances-and-contributions-to-fish-biology: intechopen.com, Retrieved 4 May, 2019

- Osmoregulation-in-fish-meaning-problems-and-controls, anatomy-and-physiology, fish: yourarticlelibrary.com, Retrieved 3 February, 2019

- Thermoregulation: fishionary.fisheries.org, Retrieved 13 July, 2019

- Fish-maintain-buoyancy-5539: animals.mom.me, Retrieved 23 February, 2019

- Fish-locomotion: iaszoology.com, Retrieved 1 April, 2019

Chapter 5

Fish Habitats and Fish Behavior

The area where all the requirements which a species of fish needs to survive are met is known as a fish habitat. The study of the response of fish towards external stimuli is studied within fish behavior. The chapter closely examines the key concepts related to fish behavior and habitat, such as migration of fish and parental behavior, to provide an extensive understanding of the subject.

Fish Habitats

Fishes require water and a wide diversity of submerged structures to provide them with the right conditions to maintain sustainable populations. The term habitat refers generally to an area (or areas) within which the requirements of all life history stages of a species are met. This may occur over a limited area, or it may be made up of a variety of different habitats over a broad landscape. An area where recruitment occurs on a scale sufficient to sustain a population may be different to the adult habitat where feeding and growth occurs. For example, Golden perch may use inundated floodplains as nursery habitats, but return to the river system for the remainder of their life cycle. Where species like Golden perch may migrate many kilometres between habitats, others, like the barred galaxies an inhabitant of small mountain streams, have a more restricted, specialised habitat.

Broad Habitat Types

There are two broad habitat types within freshwater environments, lotic habitats (flowing waters) and lentic habitats (non-flowing waters).

Lotic habitats, rivers and streams, are formed by the interactions between the landscape, geology, climate and vegetation where they are located. The intensity, frequency and duration of rainfall and the gradient combine to impose energy on the river channel. This creates structures important to fish such as pool/riffle sequences, substrate diversity, undercut banks etc.

Rivers and streams interact in three dimensions. Longitudinally, there is a general change along the channel from fast flowing upland slopes to low land, slow flowing and often meandering watercourses. Maintaining longitudinal connectivity is important for fish to be able to access different habitats at different stages of their life cycles. Laterally, watercourses connect with their floodplain often connecting with lentic habitats such as floodplain wetlands. Maintaining connectivity with the floodplain is important in maintaining the lifecycles of many fishes. Finally, there are vertical connections with the underlying bed substrate and groundwater. Changes to groundwater levels will impact on river hydrology and potentially water quality (e.g. saline intrusion).

Lentic habitats are characterised by a general lack of flow, particularly in isolated lakes. However, the most common natural lentic habitats in the Murray-Darling basin are associated with floodplains. These floodplains may take on many of the characteristics of lotic habitats during over bank flood events. Persistent floodplain wetlands form important refugia for fish when the rest of the floodplain dries up. The occurrence of lentic habitats has increased significantly since European settlement due to the construction of dams and weirs. Much of the Lower Murray resembles a lentic habitat for significant periods due to the number of locks along its length and resulting lack of flow diversity.

Habitat Diversity

Within the broad habitat types of lentic and lotic, there is a wide diversity of conditions including flow variations, substrate types, instream structures (such as log jams), depth, shading (influencing temperature etc.) and water chemistry. This results in a wide range of habitats which fish exploit at various stages in their life cycles. Koehn and Kennard (2013) make the distinction between mesohabitats (e.g. riffles and pools) and microhabitats (e.g. areas of different depth, substrate, cover etc. within a pool or riffle).

This habitat diversity is vital to inland fishes; it varies both spatially and temporally. Fish move between these habitats to access resources depending on their habitat preferences at different stages of their life cycles. Whilst some information is available on the mesohabitat preferences of many freshwater fishes (e.g adult Murray cod are most often associated with deep pools with instream cover; Southern pygmy perch, prefer slow or still waters associated with aquatic vegetation), much less is known about microhabitat preferences.

Healthy Fish Populations need Good Habitat

Since European settlement there has been significant habitat degradation throughout the Murray-Darling Basin. This has led to a decline in the numbers and distribution of native fishes. Impacts include; regulation or modification of in stream flows, physical alteration of habitats (infrastructure, river engineering), desnagging, removal of riparian vegetation, erosion and siltation caused by poor land management, changes in water quality (e.g. increased salinity), barriers to fish movement reducing access to critical habitat, introduction of alien species (e.g. Carp) etc.

The Native Fish Strategy (NFS), released in 2003, provided a long-term strategic approach to addressing the multiple causes of degraded condition of river systems throughout the Murray-Darling Basin with the aim of rehabilitating habitats and fish populations over a 50 year period. Over the last 10 years, actions under the NFS have greatly increased or knowledge of the habitat requirements of native fishes, and on-ground actions have begun to protect and rehabilitate key habitats across the Basin. For example, the demonstration reach program has promoted habitat rehabilitation through involving local communities in the rehabilitation of river reaches and demonstrating the benefits for native fish populations and river health in general.

It is important that the momentum gained through the NFS is not lost. Future rehabilitation activities should build on the experiences of the NFS and focus on protecting and rehabilitating critical fish habitats (e.g. drought refugia), maintaining connectivity between habitats and maintaining habitat diversity and quality across the Basin. Actions should continue to be underpinned with good science and an adaptive management approach.

Threats to Fish Habitat

The various threats to fish habitat are:

- Excessive removal of trees can cause soil erosion that results in increased runoff of sediment, silt, woody debris and sawdust into waterbodies, and changes in stream bank structure, affecting stream flows; and increased stream temperatures.

- Mining can cause fish habitat loss due to increased sediment loading/ toxic runoff in streams/rivers and direct removal of stream channel or bank gravel beds.

- Older agricultural practices. such as improper field fertilization that leads to nutrient flows into waterbodies, causing fish kills and nuisance algal blooms that crowd out native plants in important fish habitats; and water flow diversion for irrigation or artificial water level regulation that expose fish habitat to air.

- Residential and industrial development, such as improperly planned parking lots/roads, causes increased runoff of sediment, sand, road salt, oil, lawn fertilizers, and toxic substances into streams, rivers, and lakes.

- Inland navigation: construction of canals and unregulated dredging can change stream flows, which can either fill in or remove important habitats.

- Invasive species: many foreign invaders have colonized important fish habitats reducing water quality in these habitats and some prey on eggs and larvae of important fish.

Fish Behavior

Fish behaviour is a complicated and varied subject. As in almost all animals with a central nervous system, the nature of a response of an individual fish to stimuli from its environment depends upon the inherited characteristics of its nervous system, on what it has learned from past experience, and on the nature of the stimuli. Compared with the variety of human responses, however, that of a fish is stereotyped, not subject to much modification by "thought" or learning, and investigators must guard against anthropomorphic interpretations of fish behaviour.

Fishes perceive the world around them by the usual senses of sight, smell, hearing, touch, and taste and by special lateral line water-current detectors. In the few fishes that generate electric fields, a process that might best is called electro-location aids in perception. One or another of these senses often is emphasized at the expense of others, depending upon the fish's other adaptations. In fishes with large eyes, the sense of smell may be reduced; others, with small eyes, hunt and feed primarily by smell (such as some eels).

Specialized behaviour is primarily concerned with the three most important activities in the fish's life: feeding, reproduction, and escape from enemies. Schooling behaviour of sardines on the high seas, for instance, is largely a protective device to avoid enemies, but it is also associated with and modified by their breeding and feeding requirements. Predatory fishes are often solitary, lying in wait to dart suddenly after their prey, a kind of locomotion impossible for beaked parrot fishes, which feed on coral, swimming in small groups from one coral head to the next. In addition, some predatory fishes that inhabit pelagic environments, such as tunas, often school.

Sleep in fishes, all of which lack true eyelids, consists of a seemingly listless state in which the fish maintains its balance but moves slowly. If attacked or disturbed, most can dart away. A few kinds of fishes lie on the bottom to sleep. Most catfishes, some loaches, and some eels and electric fishes are strictly nocturnal, being active and hunting for food during the night and retiring during the day to holes, thick vegetation, or other protective parts of the environment.

Communication between members of a species or between members of two or more species often is extremely important, especially in breeding behaviour. The mode of communication may be visual, as between the small so-called cleaner fish and a large fish of a very different species. The larger fish often allows the cleaner to enter its mouth to remove gill parasites. The cleaner is recognized by its distinctive colour and actions and therefore is not eaten, even if the larger fish is normally a predator. Communication is often chemical, signals being sent by specific chemicals called pheromones.

Migration of Fishes

Migration of fish is defined as a class of movement which involves a long journey to a definite area for some purpose and impels the migrants to return to the region from which they have migrated. The purpose of the journey is breeding and feeding. Migration is a two-way journey. It includes emigration (outward journey) and immigration (return journey or inward journey). The fishes are notable for migration for the purpose of spawning. The inherent purpose of migration is not known.

Types of Migration

The various types of fish migration are:

Latitudinal Migration

This is performed by fishes like barracudas (Sphyraena) and swordfish (Xiphius) of the warm tropical seas. They migrate to north in spring and to south in autumn.

Vertical Migration

This is performed by many marine and freshwater fishes and is related to light, search of food, protection and also to spawning. The mackeral rises into the surface waters when there is a rich development of plankton. They eat on plankton and go to deep layers after feeding.

The swordfish, which normally lives in surface water move downwards to great depths to feed deep water fishes like scopelids. Many pelagic larvae of marine fishes perform diurnal vertical feeding migrations. They follow the vertical movements of their prey, the planktonic invertebrates which move down to great depth by day and rise to surface by night.

Many deep water fishes of the order Scopeliformes rises to spawn in the upper layers. Their eggs develop and often their larvae live feeding on the phytoplankton. Among freshwater fishes the clearest example of vertical spawning migration is that of the Lake Baikal Comephoridae. These fishes are viviparous and rise to surface from great depth of the lake to give birth to their larvae.

Spawning Migration

This is the migration in fishes for breeding, and so it is related to life cycle. Spawning migration is an adaptation for ensuring the most favourable conditions for the development of the eggs and the larvae. This also gives protection to early stages of fishes from predators.

There are two major types of spawning migrations. Movement from freshwater to saltwater for spawning is called catadromous migration. The reverse movement, that is, from saltwater to freshwater is termed anadromous migration.

a. Catadromous migration:

The most famous examples of catadromous fish are the eels, Anguilla rostrata, the European eel and Anguilla vulgaris, the American eel. For eel, the river serves as the feeding ground while die sea serves as the spawning ground. The stimulus for the start of migration in eel is the ripening of its gonads in rivers.

Before it enters the sea, the eye of the eel becomes enlarged, sometimes becoming four times as large as the eye of freshwater eels. Its face becomes sharper and its colouration changes the back becomes darker, while the belly changes from yellow to a silvery colour. The eel starts its migration in a wellfed condition.

During migration it spends enormous amount of energy so it becomes very thin. The migrating eel does not feed. Its alimentary canal degenerates considerably. Osmotic pressure of its blood rises and size of its swim-bladder decreases.

The eels migrate about 4500 km westwards from Europe or eastward from America and reach the breeding place in the Sargasso Sea off Bermuda. The adult die immediately after spawning in deep waters.

The fertilized eggs hatch out into transparent, ribbon like larvae, called the leptocephali. These were erroneously called glass fishes and placed in the genus Leptocephalus. They lead a pelagic life for a year or more and undergo metamorphosis into elvers (glass eels).

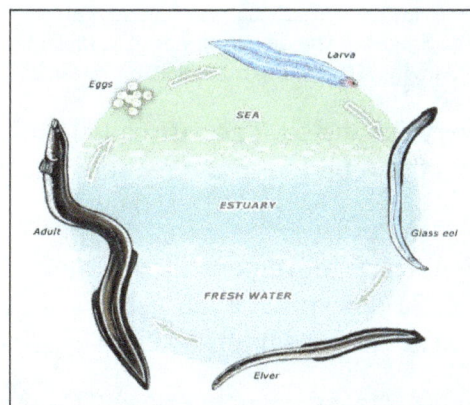

A. Leptocephali larva. B. Elver Larva. C. Adult eel.

The elvers then start ascending the rivers in shoals and grow for some years to become adult eels. The adul eel, on maturity start moving towards sea, again the cycle is repeated.

Catadromous migration is also performed by certain members of the families Galaxiidae and Gobiidae. But their migration is considerably shorter than those of the eel. They usually pass from the lower reaches of the rivers to the adjacent shallow parts of the sea.

b. Anadromous migration:

Anadromous migrations are performed by lampreys, sturgeons, salmon, some shads, cyprinoids etc. The best examples are Atlantic salmon (Salmo salar) and Pacific salmon (Oncorhynchus nerka).

In winter, the both sexes leave their feeding ground at sea to ascend the freshwater mountain streams, reaching the identical spot where they originally grew some years ago. The total distance travelled may be even upto 3600 km, at a speed of 30-40 km. per day.

They stop feeding; alimentary tract undergoes changes into a thin thread with feebly developed pyloric caecae. Change in colour and weight too occurs. Sexual dimorphism becomes evident. The male is characterized by the possession of enlarged front teeth.

After the selection of suitable spawning grounds, the salmons segregate into pairs and produce shallow saucer-like nest where spawning takes place.

Very young salmons are known as "alevins" and they remain mostly among stones. Alevins develop into next stage called "parr" and finally to adult.

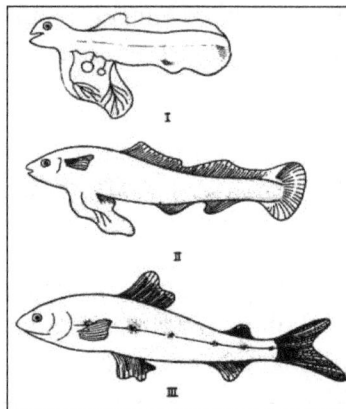

Three Stages in the development of Salmon I& II
Alevins. III 'Parr' with marks on its side.

After fertilization, salmons are very exhausted. They are called "kelts". The males seldom return to the sea. The females recover and after a period in the sea they return to breed again. This process may be repeated several times.

Some fishes do not perform significant movement like salmon. They migrate from seas to estuaries or lower reaches of river, for spawning. Such fishes are classified as fluvial anadromous (Semimigratory) fishes. Examples are many whitefishes and cyprinoids.

Many freshwater fishes leave the lakes to spawn in the river. This is called as limnodromous migration. One of the common examples of this is the whitefish Coregonus lavaretus.

Feeding Migration

This is the movement from spawning or overwintering grounds to the feeding grounds. Feeding migration can be passive or active. In many fishes the feeding migration even begins in the egg stage. It is a passive feeding migration of eggs and embryos from spawning to feeding ground.

Active feeding migration is performed by many marine fishes like cod. Horizontal feeding migration of cod comprises regular journeys, going from one good feeding ground to another.

Overwintering Migration

Overwintering and hibernation in fishes are a part of the life cycle of a fish. It is characterized by reduced activity, reduction or stoppage of food consumption, lack of food, poor oxygen condition, low temperature, drought etc. Overwintering does not occur in all fishes.

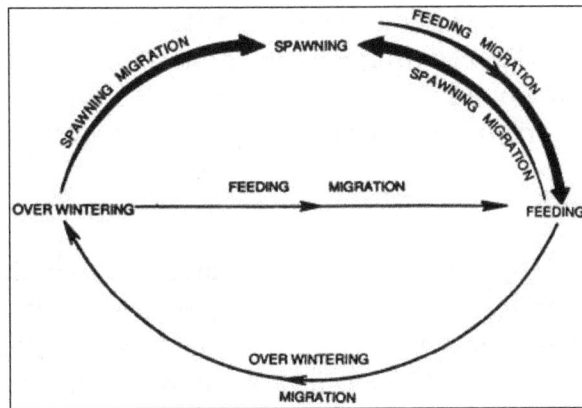

Migration cycle of various fishes.

Overwintering migration is a movement away from feeding to wintering grounds. It occurs only in those fishes which have a wintering ground.

In the wintering ground, fish is in a state of relative inactivity and reduced metabolic rate. It requires protection against predators which are common in feeding ground. Overwintering migration is performed by marine fishes like flatfishes and freshwater fishes like grass-carp.

Shoreward Migration

In this type of migration there is a temporary movement of fishes from water to land. For example, the common eel travel from one pond to another through moist meadow grass. The mud-skipper, Periophthalmus make temporary migration to land by means of modified pectoral fins. The climbing perch, Anabas migrates from water to land and even climbs trees to the height of several feet by means of the strong spines on its pelvic fins and gillcovers.

Causes for Fish Migration

Fish migration is related to several factors such as physical, chemical or biological.

1. Physical factors: The physical factors include temperature of water, rainfall, quality of water, water depths, pressure, light intensity, photoperiod, turbidity, tides and currents.

2. Chemical factors: The chemical factors include pH of water, salinity, dissolve of O_2 and CO_2, types of dissolved organic and inorganic substances, and taste of water.

3. Biological factors: The biological factors are food, attainment of sexual maturity, endocrine behaviour and competitors and predators.

Aggression in Fishes

Competition is a fact of life. It can take many forms, but biologists usually recognize two broad categories. In the first one, called exploitative or scramble competition, the contests are like races. The most food goes to the animal that eats the fastest, the best shelter is occupied by whoever reaches it first, and the largest share of eggs are fertilized by those males which produce the most sperm. There is usually little aggression displayed in such cases. However, in the second category, which is called interference or defence competition, animals fight among themselves for the right to monopolize food, to occupy alone a shelter or a territory, or to secure exclusive access to a mate. Following are some concepts and examples dealing with interference competition in fishes.

Dominance Hierarchies

Aggression allows some social fishes to sort out their relative ranks within a dominance hierarchy. Thus, when a few individuals from a social yet slightly aggressive species are placed together for the first time into a tank, a lot of nipping and chasing commonly occurs. After a while however, this aggression subsides. A pecking order has developed, every individual having figured out its place in the hierarchy. Researchers can determine the ranking of each fish by carefully observing the outcome of the initial skirmishes. The more bites an individual delivers, the more chases it initiates, and the more adversaries it wins against, then the more dominant it is. Often a linear hierarchy emerges, going from the so-called "alpha" fish at the top, to "beta" and "gamma" just below, and so on down the Greek alphabet until we reach poor "omega" at the bottom of the heap. Such a phenomenon can be observed in many salmonids, poeciliids, and centrarchids. Alternatively, the hierarchy can be despotic rather than linear. In such a case a single individual, the despot, is dominant over the other fish, which are all equally miserable. Captive eels and catfishes sometimes show this pattern.

Development of a stable and peaceful dominance hierarchy benefits everyone because fighting is energetically costly, potentially injurious, and therefore not to be done on a regular basis. However, it goes without saying that the low-ranking subordinates are not necessarily living the happiest existence. Typically their access to food is limited, as suggested by the fact that their growth rate is slower than that of dominants. For the experimenter, the challenge here resides in showing that poor growth is indeed caused directly by interference from bossy dominants, rather than poor growth and subordinate status both being caused by a third factor, such as inefficient physiology.

One way to obtain such a proof is to directly observe dominants chasing subordinates away from the best food sources, as has been done with cichlids, salmon, medaka, and brown bullhead.

One can also compare the growth rate of fish raised alone and raised in groups. The usual result here is that fish raised alone show a very consistent growth rate (little variation from one lone individual to another, indicating that their physiology is uniformly fine) whereas groups yield some fish that grow well and others that do not. The slow growers usually turn out to be those that were subordinate during behavioural interactions. Moreover – and this is the clinching experimental demonstration – the growth rate of these subordinates speeds up after dominants and other competitors are either removed from the tank or isolated behind a partition. Such a sudden improvement in growth rate would not happen if weak physiology was the original cause of slow growth.

Another sore point in the life of subordinate fish is that they seem to be more stressed. Stress probably develops because of the fear of being chased and chastised by bullies, and from having to worry more about where the next meal is going to come from. Stress reactions are often characterised by a rise in the production of certain hormones and metabolic products, and indeed the blood of subordinate fish often contains higher amounts of these substances.

The impact of stress on the life of a fish can be substantial. In some cases, it could account for at least part of the slow growth rate of subordinates. In an experiment conducted with Larry Dill at Simon Fraser University in Vancouver, Jeremy Abbott kept pairs of rainbow trout in tanks, each pair being made up of one dominant and one subordinate individual. These fish were of very similar size but this similarity did not prevent the dominant trout from nipping and charging at the subordinate. Now, at dinner time each day, Abbott separated the fish and fed them the same number of individual brine shrimp, fruit fly, or onion fly, which were all eaten by each fish. Despite the fact that food intake was the same for both dominant and subordinate, the subordinate ended up becoming smaller than the dominant in 10 out of 12 pairs. Abbott and Dill proposed that stress, along with the necessary investment in energy to repair fin damage, was the best way to explain why subordinates grew less than dominants despite their equal food intake.

Subordinates may also be forcefully relegated to less suitable habitat. In stream species such as the mottled sculpin Cottus bairdii and the longnose dace Rhinichthys cataractae, adults often can be found in deep pools while juveniles loiter in the shallows. It is tempting to conclude that the adults expel the less competitive juveniles from their preferred habitat and force them to eke out a dangerous living in the shallows, at risk from bird predation. However, we must first disprove the alternative hypothesis that juveniles stay in the shallows of their own volition, perhaps because their food requirement is different from the adults' and can best be met by foraging around weeds. Experimenters can tackle this problem by building enclosures encompassing both types of habitat within a stream, by placing adults in some of these enclosures but not in others, and finally by releasing juveniles in all enclosures. If juveniles like the shallows, they should live there irrespective of the adults' presence or absence in the pools. The results, however, are that in the absence of adults the juveniles set up shop in the deep pools. With the adults present, the juveniles end up in the shallows. So there does seem to be competitive exclusion by the more dominant adults.

And now for the ultimate evolutionary slight: the sex life of subordinates also suffers. In groups of guppies and swordtails for example, some alpha males have been reported to monopolise the area of the aquarium where females hang out, and to account for more than 80% of all copulations with them. This must leave the subordinate males fairly frustrated. Fortunately, at least in the case of guppies, subordinates can achieve some mating success by other means. They can become suave rather than strong. If they happen to have brighter body colours than dominants, or if they display more, they can garner their fair share of the mating market in spite of the dominants' attempts to suppress their sexual activity.

If subordinates remain within groups despite all of the inconveniences they must put up with, then probably the general advantages of social living – mutual defence against predators, discovery of food in greater amount than can be monopolised by a single alpha fish – outweigh the disadvantages of a limited existence next to bossy dominants. In fact, the necessity for harmonious group living may set a ceiling to the levels of aggression expressed by some fishes. A study by Anne Magurran and Benoni Seghers supports this idea. In the streams of Trinidad, some guppy populations are exposed to predators and therefore show strong shoaling behaviour, while others see few predators and shoal only loosely. When Magurran and Seghers brought these fish into the lab and placed them next to a small food patch, the strong shoalers showed little aggression among themselves while eating, whereas the loose shoalers fought with one another for better position over the food patch.8 Cohesive shoaling and high aggression don't seem to go hand in hand very well.

How stable are dominance relationships? The answer is: it depends on the species. Rainbow trout, for example, seem to remember their place in a hierarchy for a long time. Consider the following experiment by Jeremy Abbott and his co-workers. Rainbow trouts were paired up and left to establish a dominance relationship between them. If one fish was 5 % larger than the other, it always won. Then, the researchers separated the two fish of each pair and fed the subordinate in excess, so much so that it eventually became at least 15% bigger than the dominant. The two fish were reunited, and surprisingly the subordinate still cowered in front of the dominant in spite of its newly acquired size advantage. Abbott and his colleagues concluded that the risk of injury during fighting is so pronounced that trout prefer to use memory rather than renewed combat to settle contests between themselves.

On the other hand, catfish may constantly monitor the well-being of their competitors, looking for sudden weaknesses and chances to climb up the social ladder. In one experiment, John Todd, his graduate advisor John Bardach, and neurologist Jelle Atema, all at the University of Michigan in Ann Arbor, forced two yellow bullheads Ameiurus natalis to share a 190-liter aquarium. One of the two fish was clearly dominant over the other, forcing it to flee at every encounter. When this dominant was removed, kept in a separate aquarium overnight, and then returned, the submissive recognized it right away and fled from it again. However, if during its overnight leave of absence the dominant bullhead was exposed to, and beaten by, an even more dominant catfish, then upon its return to the home tank the submissive individual attacked it. The defeat during the overnight contest altered something in the former dominant (its smell, perhaps, or its bravado) and this change was immediately perceived by the former submissive, who seemed to take advantage of this weakness to stage a coup.

In the cichlid Astatotilapia (Haplochromis) burtoni, only dominant males get to occupy breeding territories. They are surrounded by younger subordinate males who often try to take over the dominants' territories. At night, all males are pale, but at first light in the morning, dominant males brighten their blue and yellow colours, and a black stripe appears near their eyes. The subordinates stay pale. However, if a dominant disappears overnight (dip-netted by an experimenter, as it happens), then within 1 h past first light a subordinate will develop the livery of a dominant male and start courting nearby females. This shows that social hierarchies are very dynamic in this species and that social opportunities can be quickly seized upon.

Territoriality

A territory can be defined as "any defended area". In fishes, territories are usually held by single individuals or by breeding pairs. The defended resource may be food, shelter, a sexual partner, spawning sites, or offspring. The defenders aimed their aggression mostly at conspecifics, although other species with similar ecological requirements (or a taste for eggs and young fish) can also be targeted.

Some territorial fishes are good at discriminating between full and partial competitors. They seem to realize the degree of overlap between their own requirements and those of other species. For example, in fishes that defend food territories on coral reefs, the more an intruding species shares the diet of the territory owner, the more often it will be attacked. Moreover, it will be challenged from a greater distance. In one typical experiment, individual fish from various species were placed in bottles. These bottles were pushed incrementally towards the shelter hole of a threespot damselfish, near the centre of its territory. The diver who pushed the bottle always retreated some distance and observed whether the bottle was attacked by the resident damsel. The results: bottled fish that were known to have a strong diet overlap with the damselfish were attacked at a distance of about 1 m from the shelter hole, whereas bottled fish that did not share the damsel's diet were either overlooked or only attacked when the bottle came within less than 0.5 m from the shelter hole. Species with intermediate levels of diet overlap were attacked at intermediate distances.

Good discrimination can also be found in black-belt cichlids, Cichlasoma maculicauda, defending their breeding territory. When there are eggs in the territory, the parents will attack egg predators and non-egg predators, but they will tolerate closer approaches by the non-egg predators before launching an attack. Moreover, egg predators that hover and seem to ogle the nest are attacked more assiduously (from a greater distance) than egg predators that just pass by. Finally, when they have fry, parents chase predators only when these predators reach the distance from which they normally lunge at the fry. This distance varies from species to species and the parents seem to know this, for they adjust their attack distance accordingly.

Territoriality is a viable strategy when two conditions are met: (1) the defended resource is sufficiently localised so that it is physically possible to defend it, and (2) there is some competition for this resource, but not too much. Let's illustrate both principles by using food as an example of defended resource.

First, the question of localised resource. Suppose a fish needs a certain daily amount of food to survive. If this amount can be found in a relatively small area, then this food clump is relatively easy to defend and territoriality becomes a suitable strategy. If the food is more spread out, territoriality can still take place but the territory will have to expand – and therefore become harder to defend. At the extreme, if food is very thinly distributed, the required territory size would have to be so large that it would be physically impossible to guard all of its boundaries. In such cases, territorial behaviour is absent.

These ideas lead to testable hypotheses. One is that territory size should increase as food becomes less abundant, but only up to a certain extent. Territory size can be estimated by linking, on a map, all of the outermost places where intruders are attacked by a given defender. It is then possible to measure the size of the area thus delimited. Food abundance, for its part, can be sampled by researchers within the territory itself (much to the dismay of the territorial fish who usually cannot defend against big human intruders – although some puny but pugnacious damselfishes have been known to try). In this way, salmon and trout have been shown to indeed occupy larger territories when food is less abundant.

With a more experimental touch, Mark Hixon, of the University of California at Santa Barbara, found territorial male black surfperch Embiotoca jacksoni, a coral-dwelling species, and decreased the amount of food accessible to them. He did this by covering portions of the coral reef with swatches of nylon netting through which the fish could not pass. Unfazed, the surfperch reacted by increasing the size of their territories and appropriating neighbouring patches of untouched coral. However, when the reduction in food supply was very great, as occurred when overgrazing sea urchins invaded the area, the surfperch abandoned their territories rather than try to expand their boundaries.

The disappearance of territorial behaviour when food becomes too thinly distributed has also been documented in controlled lab experiments. In most cases, automatic feeders were set up over a very large aquarium containing many fish. The feeders were programmed so that only one of them dispensed all the food or each one of them gave a little bit of food. The prediction was that fish would become territorial in the first instance but not in the second and this is indeed what happened. When only one feeder was activated, the dominant fish in a group soon started to defend the area around that feeder. When all the feeders were activated, it became impossible to defend all of them. The group therefore spread out and swam all over the tank. Very few individuals, if any at all, bothered to defend a given area. This was observed in studies with medaka, pygmy sunfish, salmon, and juvenile cichlids.

In the wild, feeding territories are seldom observed in freshwater habitats, but they are fairly ubiquitous over coral reefs. One possible explanation for this state of affairs is that coral reefs offer a richer supply of food – their rate of production has been estimated to be 10 times as high as that of an average lake or stream. Therefore, over coral reefs, food can be concentrated in one area sufficiently small to be successfully defended. Conversely, the relative lack of territoriality in freshwater habitats could be linked to the greater dispersion of food sources. This is not to say that freshwater species are incapable of establishing territories when conditions are right, in other words when resources suddenly become clumped. The species mentioned in the preceding paragraph, where dominant individuals defended territories around single feeders in the lab, were all freshwater species.

Now, on to the question of competition level. If food is superabundant everywhere, so much so that everyone gets to eat to their heart's content and no competition exists, then obviously there is no point in establishing a territory. At the other extreme, if the number of competitors and intruders is so high that a single fish cannot defend even the smallest of territories against everyone, then obviously there is no point in territoriality. It is at intermediate levels of competition that territoriality becomes a viable prospect.

These ideas can also be tested in the lab. One can set up a great number of feeders over a tank, and switch from a situation where only a few feeders are giving food (territories will probably be defended around them) to an all-you-can-eat buffet where all the feeders are offering a lot of food. In such a tank of plenty, harmony should reign supreme and aggression should subside. In medaka, this is what happens. Interestingly, if some of the medaka get paranoiac and persist in their belligerent ways even though food is superabundant, they turn out to grow more slowly, probably because of all the energy they needlessly spend in chases and threats directed at others.

Intruder pressure, for its part, can be experimentally intensified simply by adding fish to a tank. A number of species have thus been shown to have a harder time maintaining a territory when competitor density is very high. One of the first reactions of territory owners when intruders become too numerous is to decrease the size of their territory to make it easier to defend against this onslaught of trespassers. Later, if competitors (or alternately, territory neighbours) are experimentally removed, the remaining residents expand their domain back to what they see as an ideal size.

Note that topography can also influence territory size. In habitats that are structurally complex, with lots of rocks or plants (some of which may have been added by curious experimenters), territorial defense is more difficult because boundaries cannot be visually monitored all at once. This results in smaller territories, and in more fish cohabitating next to one another. Topography may also provide natural landmarks that act as bastions for territorial defence, and if such landmarks are abundant, the tendency to use them as borders may lead to smaller territories.

If a fight erupts between two fish, can we predict which one will be the winner? The answer is yes if there is a big size difference between the contestants. As one might expect, big fish have the upper hand. One example of this comes from Indiana University, where William Rowland kept a large number of three-spined sticklebacks in stock tanks. These fish were somewhat crowded in a bare environment and could not establish territories. Rowland dipnetted various males from the stock tanks and weighed them. To put these males in a fighting mood, he placed them into individual aquaria and let them establish breeding territories. Then he picked two of these territorial males at random and moved them together to yet another aquarium, one that was unfamiliar to both of them, although it still looked like their own. The two males soon faced each other and initiated a fight. After much spine-erecting, head-to-tail chasing, and biting, the loser declared itself by breaking off the fight and cowering in a corner. Rowland staged 31 such encounters in which a clear winner emerged, and he found that the heaviest male was victorious in 22, or 71%, of them. Statistical methods revealed that the greater the weight difference was between contestants, the greater the chances that the heavier fish would end up winning. A weight difference of 15% practically guaranteed victory for the heavies.

In the experiment above, it was important to stage the contest in a neutral arena unfamiliar to both fish. If the contest had taken place in the home tank of one of the two sticklebacks, the territory owner would have held a much greater probability of winning than might be inferred simply from its body size. Being in one's own territory seems to confer more confidence, or perhaps a greater realisation of what is at stake for the owner. Sports fans call this the home turf advantage. Ethologists prefer to speak of a "prior residency effect". Within certain limits, the prior residency effect is enough to prevent large intruders from usurping the territory of smaller residents.

The duration of prior residency may also have an influence: in brown trout Salmo trutta, longer-term residents (4 days) outperform shorter-term residents (2 days) in the defense of their territory.

Another factor that may influence the outcome of a fight between two closely-matched fish is a prior experience of submissiveness. A fish that has just lost a contest is more likely to give up during the next fight as well, even when this second fight is against a new adversary. It is as if the first setback created a general losing state of mind. In sticklebacks, this lingering "loser effect" can last up to 6 h.

As an example, we can look at the work of Theo Bakker and his colleagues at the University of Leiden in the Netherlands. These researchers kept a great number of territorial male sticklebacks in individual tanks. They chose a few individuals at random and subjected them to a losing experience by dropping them into the tank of another male: this other male beat them up, taking advantage of the prior residency effect since he was in his home territory. Other males were also chosen at random and they experienced a win by having another male dropped into their own tank (for them, the prior residency effect worked in their favour). Three or six hours later, these respective losers and winners each met an inexperienced male within the confines of a neutral arena unfamiliar to all of them. If it was not for their previous experience, we would expect the previous losers as well as the previous winners to dominate this new encounter on only half of all tests since all of these fish were chosen at random. Previous winners indeed won only half of the time, indicating that their previous winning experience did not make them stronger. But a different picture emerged in the case of the previous losers: none of them won a single fight when this fight was held 3 h past their first debacle. Even after 6 h, the previous losers won on only 20% of all tests.

Similar results have been obtained with blue gourami, paradise fish, green swordtail, and pumpkinseed sunfish. In the case of the blue gourami and the pumpkinseed sunfish, it seems that a previous winning experience can instill confidence and help a combatant win its next encounter. This winner effect does not last very long in the sunfish – no more than 1 h – but it does persist for at least 3 days in the gourami.There is also the Mangrove rivulus Rivulus marmoratus, where both a winner and a loser effect exist for about 2 days. Maybe prior experience fine-tunes the information a fish has about its own fighting ability. Or maybe the fight alters hormone production differently in winners and losers, with an impact on their willingness to fight again. (This hormonal hypothesis has been invoked to explain another short-term winner-loser effect: in the Mozambique tilapia Oreochromis mossambicus, 15 minutes after a fight, winners court females more readily, for a longer time, and with more courtship sounds than losers, even when each male is alone with the female).

Fighters are more successful when they can first impress their opponents with signs of their good health and good growth. Sometimes, these signs are the same ones that are used to woo females. In swordtails for example, males possess an elongated lower tail section that resembles a sword. Females prefer to mate with males who have longer tails; interestingly, males with longer tails also win more fights, even when matched with opponents of similar body size. It is possible to attach plastic extensions to the tail of a given male, and all of a sudden this male starts to win more fights than he used to do, probably because his big sword incites his adversaries to back down. In the same vein, male sticklebacks with brighter red throats attract more females, and also win more fights against other males. Intimidation is probably involved rather than actual fighting ability, because when fights are staged under blue light (which makes the throat appear black instead of red), males with brighter red throats do not win more fights any more.

Intimidation is behind all of the ritualized displays that are performed by both contestants at the beginning of a fight. In fishes, such displays include booming sounds, water-displacing tail beats, fin erection, gill cover spreads, head shakes, body twists, lateral displays that reveal the full size of the body, colour changes, exposure of brightly coloured body parts, and intricate swimming manoeuvres. These actions are meant to signal fighting ability and to encourage opponents to give up.

Other determinants of dominance during fights include stamina and motivation. Escalated fights may last for a long time (a half-hour is not uncommon in some species) and stamina would prove an asset in such a situation. Consider the work of Francis Neat and co-workers on the redbelly tilapia Tilapia zillii. These researchers found that losers of territorial fights harboured more lactate within their muscle than winners did immediately following the fight. Lactate is a metabolic by-product that can cause fatigue. So the vanquished fish may have lost because they were the first ones to get tired out.

Motivation, or "fighting spirit", could also characterise good combatants. In another study on redbelly tilapia by Neat, small males sometimes won over larger ones. These smaller winners were more aggressive during the fights and inflicted more bites. They also had larger gonads, indicating that they were more ready to spawn, and therefore perhaps more inclined to defend their breeding territory. Being closer to spawning may also explain why breeding pairs of convict cichlids that have been together for a longer rime (96 h versus 48 h) fight more successfully for breeding sites.

In some species, motivation to fight may also be influenced by a "priming" effect. If a male fighting fish who has just witnessed a combat between two other males is allowed to interact with a male who has just seen two other fish not fighting, the former usually behaves more aggressively than the latter. It's as if viewing a fight put the male in a fighting mood. In the same vein, three-spot gouramis who have learned to associate the appearance of a red light with the imminent arrival of an opponent win more fights when they are forewarned by the red light.

Fish Spawning

Spawn is used to describe the process of how fish mate and reproduce. It is very different from the usual ways of reproduction that mammals have. Fish can reproduce in one of three ways. They can reproduce by livebearing, spawning, or by themselves. Livebearing is similar to the way mammals reproduce. Some fish can even switch their gender which means they are capable doing everything on their own. The most common way for fish to reproduce is by spawning. Spawning is when the female lays her eggs in the water. The male will then come and fertilize as many eggs as he can.

Whether or not a fish lays eggs depends on the type of fish. Some fish birth live fish, while others do indeed lay eggs. A female laying their eggs for the males to fertilize is the most common form of reproduction among fish. The female fish can leave anywhere from a few hundred to thousands of eggs.

Most fish follow certain spawning patterns, although that pattern changes for each specific type of fish. Many change location to lay their eggs or give birth in specific places or by certain underwater terrain for protection. Some fish, like salmon, are almost always caught during their spawning season.

Spawning Season

Although the concept for spawning is the same for all the fish, there are some differences for different types of fish. Those differences include the time of the year, the age of the fish participating, and their behaviors when spawning. Different types of fish have different strategies when it comes time to spawn and they have to worry about different threats.

Catfish Spawning

Catfish tend to spawn during the spring and summer months. They wait for the water to warm to an optimal 70-84 degrees. Catfish are also considered the most difficult fish to catch while they are spawning. Part of that is due to the fact that they are nesters. They lay their eggs in dark, secluded cavities they find. These cavities can be found between rocks, in hollow logs, or burrows in banks.

After the female picks a nesting location, she lays her eggs. A male will then come and fertilize the eggs. After he fertilizes the eggs, the females leave and let the males protect them. The male fish keeps predators away and fans the eggs so they do not get buried by sediments. After about a week the eggs will hatch.

Bass Spawning

When bass are ready to spawn, it only happens in the spring. They wait for the waters to be anywhere between 70-80 degrees. Even though bass spawning season only take place in the spring, they have three phases that they go through.

The pre-spawn is when bass begin eating much more. They increase the amount of food they eat because they know they will not eat while they are actually spawning.

While bass are spawning they become very aggressive. Species that they used to feed on will now be the ones they attack just to keep their eggs safe.

During the post-spawn the eggs have hatched. The males will guard the newly hatched fish while the females leave to rest in deeper water.

Salmon Spawning

If you've ever seen footage of salmon hopping upstream, you've seen them on their way to spawn. Salmon migrate upstream to lay their eggs. They swim back to the same area where they hatched to spawn,and many die during or after the process.

Salmon's journey upstream to spawn is called the salmon run, and it's important to know because it is the best time to catch salmon. Salmon are born in freshwater rivers and streams and return there to spawn, but they spend most of their life in the ocean. They can change habitat, and even change colors when spawning – many turn a bright red. Most anglers catch salmon during their migration on the way to spawn.

Salmon usually spawn in the spring through the summer, and there are resources that track salmon runs by specific area or even body of water.

Betta Fish Spawning

Spawning bettas can be very difficult. The breeding pair needs to carefully match because the males tend to be aggressive. The best time for bettas to spawn is when they are young. Since they are fish that are typically pets, the time of year does not matter. For an ideal offspring, the male and female should spawn before they are older than a year old.

The females tend to be the picky ones when selected a male to fertilize the eggs. Some things they look for in a suitable male are his energy levels, coloration, and fins.

The male will start to build a bubble nest while also showing off for the female. Once he is done building the nest the female will make sure it meets her standards or else she will destroy it. After they begin to mate, it will be a couple hours of them chasing after each other and biting.

After some time, she will begin to release her eggs. The male will then collect the eggs and put them into the bubble nest one at a time. The male will then be in charge of protecting the nest and he will attack anything he sees as a threat.

Goldfish Spawning

Goldfish are able to spawn when they are one or two years old. They usually spawn once a month through spring and summer months. They prefer the water to be 70 degrees while spawning during the day. While spawning, the male will chase the female until she releases her eggs so he can fertilize them.

Parental Behavior

The important aspects of parental care by the fishes are:

Additional Strategies

Many types of additional structures are present on the eggs to provide protection to them. In elasmobranches and holocephalans the shell gland of oviduct secretes a horny case around the egg. These cases protect them against predation and damage, and to certain extent help in osmotic control.

An egg of some fishes like sculpin, trouts and minnows are coated with adhesive secretion, which keeps them held together and also helps them to attach to some objects so that they are prevented from being washed away.

Selection of Spawning Site

Spawning at sheltered place is one of the most primitive types of protection afforded by the parents. Migratory fishes migrate to long distance in search of suitable place for spawning.

The anadromous fishes like Petromyzone marinus, Acipenser, Hilsa and Salmo migrate extensively to spawn in suitable waters. The potadromous fishes like carps and trouts travel long distance within water in search of suitable spawning site and then return to their feeding grounds.

Parental Care by Nest Building

The most primitive is the habit of nest building. Many fish species build nest in one form or another, whether it is a simple pit dug into the gravel or the elaborate bubble nest. No special breeding setup is needed. When ready to spawn the fish constructs a nest by blowing bubbles, often using vegetation to anchor the nest.

The male keeps the nest intact and keeps a close eye on the eggs. The female should be removed after spawning. Care is needed to raise the fry and the tank should have a glass cover to help keep the nest moist and warm. The Gouramis, Anabantids and some catfishes are the most common examples of this type of spawners.

The nest may be a simple trough or hollow cleaned out in the sandy bottom of a water body, as in case of salmon. Often one of the parents, usually the male, guards the eggs and undertakes the responsibility of incubating eggs and of rearing the fry till they are able to take care by themselves.

The Lepidosiren paradoxa of South America constructs nest in the form of closed burrow in a swampy soil. It has been reported that during the breeding season male fish develops filamentous structures on their pelvic fins. Since these filaments are highly vascular, they secrete oxygen into water. According to some other workers, it is suggested that the filamentous gills supplement the oxygen requirement of the male fish, which during the time reduces its frequency to visit the surface.

The lungfish of Africa, the Lapidosiren scoops out a hallow depression in the mud of swamp, rich in aquatic weeds and grasses, which afford protection. The female lays eggs in a hollow nest. After that the male fish undertakes whole responsibility for nest building and subsequent care of the eggs and young ones. It swims around, chasing the predators and aerates the water in the nest for the eggs and young larvae.

The guarding parents usually stops feeding and utilize their body reserves during the period of care. involvement of male parent in care and protection of young ones to make female parent free from burden of care so that they may feed properly and grow normally to maintain their fertility.

The bullhead, Ictalurus natalis either utilizes old tunnels of rodents or prepares a hole. The male ventilates and guards thousands of eggs. The male looks after the young ones even after hatching till they grow up to two inches in size.

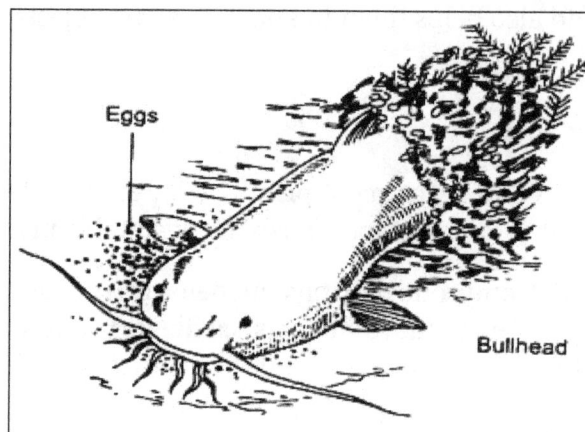

Parental care by male Bull head.

Female Trout, Tilapia and the Barbus for prepare their nest by scooping in gravel bottom, although they have different egg laying habits and different breeding seasons.

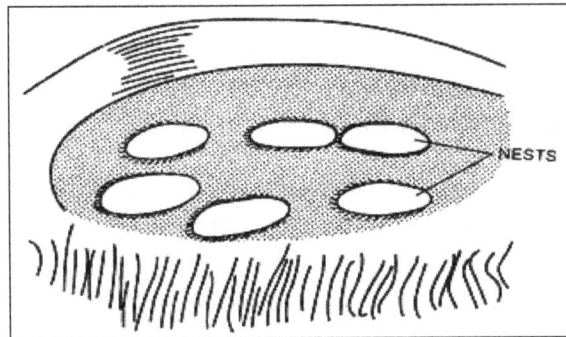

A drained pond showing nests of tilapia.

The African Osteoglossid (Clupisudis) makes a nest by cleaning a space in the aquatic vegetation. The nest is built in about two inches of water and it is about four inches across. The wall of the nest is several inches in thickness, made up of stems of grasses. The smooth bare ground constitutes the floor of the nest.

Females of Johney darter (Etheostomal) of Eastern North America build the nest in the form of shallow depression. The eggs laid are coated with mucus, so that they can stick on to stones lying around the nest. The males of the three-spined stickleback, Gasteroteus, construct a much more elaborate nest.

Their kidney secretes a special sticky and glue-like product. With the help of this glue a male bind bits of plants and prepares the nest. Then male makes searches for a female, courts her and invites her to the nest, where she laid the eggs which he fertilizes. Soon after he again searches another female and the processes repeated till the eggs fill the nest. The male then undertakes the responsibility of guarding the eggs till they hatch. Ophiocephalus prefers weedy ponds to deposit their eggs.

The male protects the eggs and later leads the youngs in school, till they grow about two inches long. Male of giant kelp fish, Heterostichus rostratus, guards the nest carrying fertilized eggs. During the period of care the male gets aggressive against all predators.

Gymnarchus (Mormyrid) forms a large floating nest, which lies projected many inches above the water surface. The fishes enter in the nest from the end, which remains 6″ below the water surface.

Male of the Siamese fighting fish (Betta splendens) of family Anabantidae, builds nest made of air bubbles. They blow air bubbles in mucus of mouth, which are accumulated on the under-surface of some objects, floating in water. When enough bubbles are accumulated, courtship follows, and the eggs are fertilized. The male then relaxes and takes up position below the female. It picks up fertilized eggs in the mouth and swims upwards to adhere eggs to the nest by mucus. Males look after the nest, repair the nest and protect from predators. In these species the protective instinct in males is so strong that they often fight to enemies for defence of eggs till death.

Amia (North American) forms circular nests. They construct nests in swampy area of a lake, rich in aquatic plants. One or many females deposit their eggs at the bottom of the nest. The males then guard the nest. After hatching the young fishes leave the nest and swim in groups, which are guarded by males by encircling around them.

Lepomis gibbosus, pumpkin seed fish scoops out shallow depressions at the shores. The males construct the nest and escort the females for egg laying. A female after laying eggs is, driven away by the male, and then searches for another female for depositing the eggs. Thus the nest contains eggs of many females. The males protect the nest until hatching takes place.

Male and female Amiurus nebulosus, a species of North American catfish, construct small crude nest in the form of a depression excavated in mud, along the bank of rivers or under any object in water. When spawning takes place, the male parent undertakes the protection of nest and later takes the young ones in school till they are shifted to safe water of shore.

Similar type of nest building habit is found in North American sunfish (Centrarchidae). The males of this fish scoop out a shallow and circular nest in the gravel bottom. The males prepare the nest and remove large stones, and only a fine sand layer is left. The females lay eggs, which are stick to sand. The male looks after the eggs until hatching takes place.

Male sunfish guarding eggs in nest.

Egg Depositors

In this case, the eggs are either laid on a flat surface, like a stone or plant leaf or even individually placed among fine leafed plants like Java moss. The parents usually form pairs and guard the eggs and fry from all danger. The Cichlids are the best-known species for this. Some catfish and rainbow fish are also egg depositors.

The set-up for these fishes will vary with the species, but usually a flat stone, broadleaf plant, cave or a broken flowerpot are to be provided. Sometimes you can remove the item that the eggs have been laid on to a separate hatching tank. Paradise fish provides care to their eggs, which are lighter than water hence rise up and attach to bubbles in nest. Either of parents collects those eggs in mouth if falling down.

Parental Care by other Means

One of the notable examples of parental care is found in a small carp (Rhodeus) commonly called Bitterling, inhabiting the rivers of Central Europe. These fishes do not make nest, but discover tailor-made nest in the shells of the freshwater mussel.

The female fish draws out its oviduct to form a long tube just like an insect's ovipositor. With the help of this tube the females deposit their eggs in the bivalve shell, a safe place for eggs.

Inside the shell the eggs undergo development and are aerated by respiratory current, which flows over the gills of mussel. It is observed that this period extends up to a month. It is amazing to note that the breeding season of the mussel coincides with the breeding season of the Rhodius fish. The glochidia larvae are temporary parasite on the fish and get dispersed wherever the fish swim about.

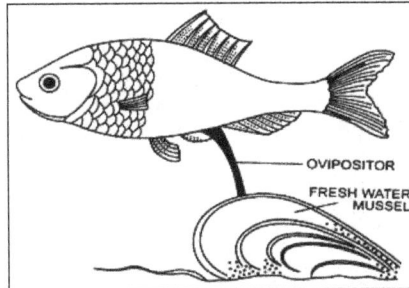

Female bitterling depositing eggs in the gill
chamber of freshwater mussel.

Another interesting type of parental care is seen in the Brazilian catfish (Platystacus). The belly of female of this species becomes soft and spongy during the breeding season. The female lies on the fertilized eggs, which get attached to the skin by small stalked cup-like structure. The cup is vascular and nourishes the developing embryo until they hatch.

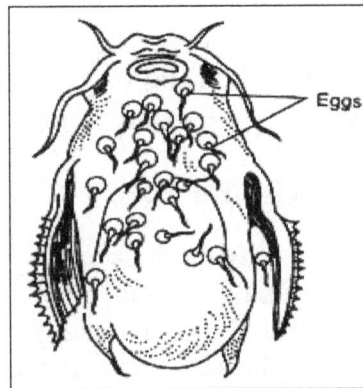

Parental care by female obstetrical cat fish.

Some species provides parental care by modification of certain structures. The lower lip of male of a Brazilian fish enlarges in the form of a pouch. The fertilized eggs are incubated in this pouch. However, mouth itself is used as an oral incubator in two catfishes, viz., Bagre marinus and Geleichthys jelis.

The males of these species carries in his mouth 10-30 developing eggs. The African fish Tilapia mossambica is known as 'mouth brooder' as hatchlings come in the oral cavity at the time of danger.

Tilapia mossambica: the young one taking shelter in
the female buccal cavity when danger apprehended.

The Butterfish (Pholis) makes a ball of its spawn and roll around carry in their mouth to protect the eggs. Both the parents carry out this movement alternately.

Butter fish pholis coiled round a mass of eggs.

Parental Care by Mouthbreeders

The females usually lay their eggs on a flat surface where the male then fertilizes them. After fertilization the female picks up the eggs and incubates them in her mouth. Even after hatching the fry will return to the safety of their mother's mouth if danger is near.

Brood numbers are usually small, since by the time the fry are released they are well formed and losses are minimal. The best-known mouthbreeders are the African lake cichlids.

The female fish keeps the eggs in her buccal cavity and provides shelters to the young ones. The males of marine catfish (Aridae) collect the fertilized eggs in the mouth and keep them till their complete development takes place. During this period they do not feed.

Parental Care by Modifying Special Parts

Some fishes attribute parental care by modifying their organs in special manner. Male of Florida pipefish (Syngnanthus) develops a pair of flaps on underside of body. During the breeding season these flaps fuse to form a brood pouch. The female extends its oviduct and deposits the eggs into the brood pouch. The development takes place into brood pouch.

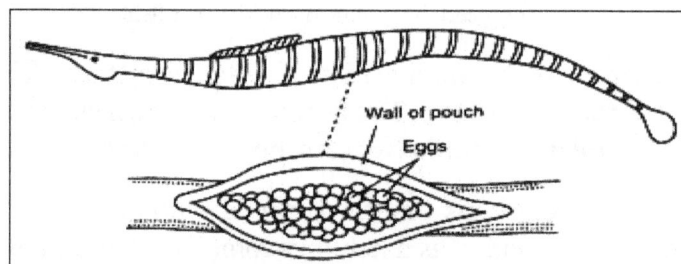

Male pipefish sygnathus and its brood pouch dissected and opened to show eggs.

The Australian kurtus (Kurtoidei) offers parental care to the fertilized eggs in a different manner. The male fish produces its dorsal fin and forwards into a spinous hook, to which the eggs adhere to in clusters. In the Hippocampus (Sea horse), the male possesses brood pouch beneath the tail.

At the time of breeding season this pouch becomes thick walled and receives many blood vessels. The cloaca of the female becomes extended to form a genital papilla, which serves as an intermittent organ for the transfer of eggs to the brood pouch of male. When the eggs hatch into embryos, the vascular wall of the pouch nourishes them. At a time at least 6-7 miniature seahorse may be accommodated in the pouch.

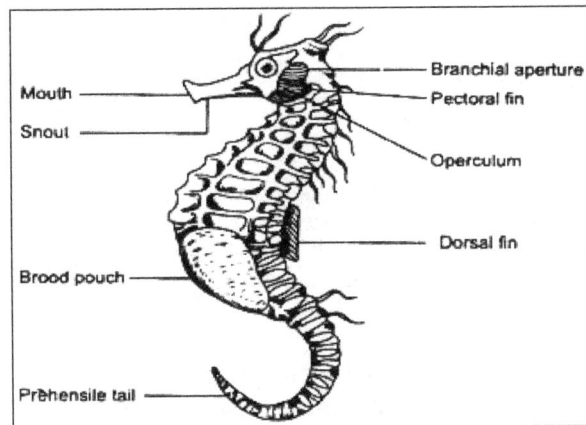

Parental care a male hippocampous.

Parental Care by Egg Buriars

The annual Killifish are known for this method of parental care. As the pools where they live dry out, the fishes spawn pressing their eggs into the substrate. The pools dry out completely and the adults die, but the eggs remain in the dried mud. When the rains return and the pool refills, the eggs hatch and the cycle is repeated. Killifish eggs can stay viable for many years in the dried out mud.

Influence of Endocrine Glands on Parental Behavior

It has been reported that more than one hormone regulate the reproductive behaviour of the fishes. prolactin hormone' accelerates the secretory activity of buccal epithelium of Betta at the time of breeding season.

Thus the fish becomes enabling to secrete air bubbles in mucus for building of nest. In another fish Symphyrodon discus hormone prolactin, which is originated from testosterone and is called as paralactin, activates mucus secreting epidermal cells.

We know that mentioned that nest building behaviour of Betta is controlled by testosterone, which activates only in presence of hormone prolactin. In Lepomis, Acquidens and Pterophyllum the fanning behaviour is induced by prolactin in the presence of gonadal steroids. In case of Hippocampus it is prolactin that is responsible for proliferation of epithelium lining of the brood pouch.

References

- Fish-habitat: finterest.com.au, Retrieved 21 August, 2019

- FishHabFactSheet-FishHabitat0110, FishFactSheets: seagrant.sunysb.edu, Retrieved 25 March, 2019

- Behaviour, fish, animal: britannica.com, Retrieved 2 August, 2019

- Migration-in-fishes, fisheries: biologydiscussion.com, Retrieved 21 May, 2019

- Migration-of-fishes-phylum-chordata, fish, fisheries: biologydiscussion.com, Retrieved 25 July, 2019

- 9-important-aspects-of-parental-care-by-the-fishes, anatomy-and-physiology, fish: yourarticlelibrary.com, Retrieved 5 April, 2019

Permissions

All chapters in this book are published with permission under the Creative Commons Attribution Share Alike License or equivalent. Every chapter published in this book has been scrutinized by our experts. Their significance has been extensively debated. The topics covered herein carry significant information for a comprehensive understanding. They may even be implemented as practical applications or may be referred to as a beginning point for further studies.

We would like to thank the editorial team for lending their expertise to make the book truly unique. They have played a crucial role in the development of this book. Without their invaluable contributions this book wouldn't have been possible. They have made vital efforts to compile up to date information on the varied aspects of this subject to make this book a valuable addition to the collection of many professionals and students.

This book was conceptualized with the vision of imparting up-to-date and integrated information in this field. To ensure the same, a matchless editorial board was set up. Every individual on the board went through rigorous rounds of assessment to prove their worth. After which they invested a large part of their time researching and compiling the most relevant data for our readers.

The editorial board has been involved in producing this book since its inception. They have spent rigorous hours researching and exploring the diverse topics which have resulted in the successful publishing of this book. They have passed on their knowledge of decades through this book. To expedite this challenging task, the publisher supported the team at every step. A small team of assistant editors was also appointed to further simplify the editing procedure and attain best results for the readers.

Apart from the editorial board, the designing team has also invested a significant amount of their time in understanding the subject and creating the most relevant covers. They scrutinized every image to scout for the most suitable representation of the subject and create an appropriate cover for the book.

The publishing team has been an ardent support to the editorial, designing and production team. Their endless efforts to recruit the best for this project, has resulted in the accomplishment of this book. They are a veteran in the field of academics and their pool of knowledge is as vast as their experience in printing. Their expertise and guidance has proved useful at every step. Their uncompromising quality standards have made this book an exceptional effort. Their encouragement from time to time has been an inspiration for everyone.

The publisher and the editorial board hope that this book will prove to be a valuable piece of knowledge for students, practitioners and scholars across the globe.

Index

www.ingramcontent.com/pod-product-compliance
Lightning Source LLC
Chambersburg PA
CBHW082050190326
41458CB00010B/3503